Riccardo Tesser, Vincenzo Russo

Advanced Reactor Modeling with MATLAB

Also of Interest

MATLAB® Programming.
Mathematical Problem Solutions
Dingyü Xue, 2020
ISBN 978-3-11-066356-3, e-ISBN 978-3-11-066695-3

Basic Process Engineering Control
Paul Serban Agachi, Mircea Vasile Cristea, Emmanuel Pax Makhura, 2020
ISBN 978-3-11-064789-1, e-ISBN 978-3-11-064793-8

Chemical Reaction Engineering.
A Computer-Aided Approach
Salmi, Wärnå, Hernández Carucci, de Araújo Filho, 2020
ISBN 978-3-11-061145-8, e-ISBN 978-3-11-061160-1

Computational Chemistry Methods.
Applications
Ponnadurai Ramasami (Ed.), 2020
ISBN 978-3-11-062906-4, e-ISBN 978-3-11-063162-3

Theoretical and Computational Chemistry.
Applications in Industry, Pharma, and Materials Science
Iwona Gulaczyk, Bartosz Tylkowski (Ed.), 2021
ISBN 978-3-11-067815-4, e-ISBN 978-3-11-067821-5

Riccardo Tesser, Vincenzo Russo

Advanced Reactor Modeling with MATLAB

Case Studies with Solved Examples

DE GRUYTER

Authors
Prof. Riccardo Tesser
Department of Chemical Sciences
Università degli Studi di Napoli Federico II
Monte Sant'Angelo, Via Cintia
80126 Napoli
Italy
riccardo.tesser@unina.it

Prof. Vincenzo Russo
Department of Chemical Sciences
Università degli Studi di Napoli Federico II
Monte Sant'Angelo, Via Cintia
80126 Napoli
Italy
v.russo@unina.it

ISBN 978-3-11-063219-4
e-ISBN (PDF) 978-3-11-063292-7
e-ISBN (EPUB) 978-3-11-063231-6

Library of Congress Control Number: 2020944106

Bibliographic information published by the Deutsche Nationalbibliothek
The Deutsche Nationalbibliothek lists this publication in the Deutsche Nationalbibliografie; detailed bibliographic data are available on the Internet at http://dnb.dnb.de.

Cover image: Carmelina Rossano
Typesetting: Integra Software Services Pvt. Ltd.
Printing and binding: CPI books GmbH, Leck

www.degruyter.com

The purpose of computation is insight, not numbers
Richard Wesley Hamming

Preface

This book is a result of a long and fruitful cooperation between the authors who devoted time and experience to introduce Matlab in bachelor's and master's degree courses in industrial chemistry at the University of Naples Federico II, in the frame of the Chemical Sciences Department. The efforts were able to extend the approach to research in cooperation with industrial and academic partners, leading to a wide number of scientific publications.

The presented approach is oriented on the simulation of chemical reactors using Matlab software, describing in detail how to implement the mass and heat balance equations. The idea is to teach the audience how to use the built-in Matlab functions in a correct and efficient manner, developing numeric tools in describing experimental data collected using both lab- and pilot plant-scale reactors.

Many books are available on the market regarding chemical reactors description, ranging from basic to advanced approaches on the topic. Notwithstanding this very large published material, very few of them offer the reader a modern approach to reactor description and modeling. Our idea is based on the use of a widely spread numerical environment, Matlab, to develop categorized groups of generalized codes. Examples of application will be reported for every reactor treated in this book. For practical reasons, no specific molecules will be considered in describing chemical reactions, preferring a more general schematization (e.g., $A \rightarrow B$). This is necessary to keep the codes compact and help readers in adapting the models to their real applications. The codes presented in the book aim both to simulate almost each type of chemical reactor and to be easily extended and adapted by the reader to their own specific problem; thus, it represents a useful tool for professionals working in industry and academy.

The main difference, and contemporarily the strong point of this book, is the general validity of the developed codes when problems in reactor simulation arise. The codes are not dedicated to specific chemical systems but are of general validity and are not only demonstrative examples but, instead, also used in reality. Finally, the modular approach adopted in the book could be exploited by advanced readers to combine and further develop the codes into a more complex process simulation.

The concluding chapter is devoted to the presentation of real examples of application, choosing modern examples of chemical reaction engineering to show the high potentialities of the presented approaches. In this case, the codes are not provided for lengthy reasons; therefore, both the theory and the calculation strategies given by general flow charts are given to help readers develop their own code.

The authors are grateful to Mathworks® for constantly developing and upgrading the software. Authors acknowledge the colleagues of the industrial chemistry sector at the University of Naples Federico II (Prof. Martino Di Serio and Prof. Elio

https://doi.org/10.1515/9783110632927-202

Santacesaria) and of the chemical reaction engineering sector at Åbo Akademi (Prof. Tapio Salmi and Prof. Johan Wärnå) for the fruitful discussions that helped in realization of this book. The author thank Carmelina Rossano for creating the original version of the artwork of the book cover.

Napoli, 24th June 2020

In faith,
Prof. Riccardo Tesser, Prof. Vincenzo Russo

Contents

Chapter 1
Introduction on chemical reactors modeling

1.1 Chemical reactors modeling

Modeling of chemical reactors is a specific sector of chemical reaction engineering (CRE), dedicated to the description of the chemical and physical phenomena occurring in a reactor. The number of reactor types is rather large; thus, each reactor needs a specific approach to be modeled [1–4].

Focusing on the phenomena occurring in a chemical reactor, it is possible to develop a reactor model starting from a generic classification of the main types of these devices:

- Batch and fed batch
- Steady-state continuous tubular
  - Plug flow, laminar flow, nonideal flow
  - Packed bed, trickle bed
- Dynamic continuous tubular
- Continuous steady-state/dynamic stirred tank

Independent of the specific reactor type, the physical phenomena occurring can be always done writing down the conservation laws for mass, heat and momentum balances, that means defining the terms of: (i) reaction and (ii) mass/heat/momentum transfer.

The derivation of a unified approach to define differential mass/heat/momentum balances are given as follows [1]:

$$\frac{\partial C_i}{\partial t} = -\nabla(C_i u) - \nabla J_i + \sum_j \nu_{ij} r_j \tag{1.1}$$

$$\sum_i M_i C_i c_{p,i}\left(\frac{\partial T}{\partial t} + u\nabla T\right) = \nabla(\lambda\nabla T) + \sum_j (-\Delta H_j) r_j + \sum_i J_i \nabla H_i + Q_{\text{rad}} \tag{1.2}$$

$$\frac{\partial(\rho_f u)}{\partial t} + \nabla(\rho_f u u) = -\nabla P - \nabla s + \rho_f g \tag{1.3}$$

All the terms could be grouped as follows, showing a deep rationalization of the approach:

Accumulation:	$\partial C_i/\partial t, \partial T/\partial t, \partial(\rho_f u)/\partial t$
Convective flow:	$\nabla(C_i u), u\nabla T, \nabla(\rho_f u u)$
Chemical reaction:	$\sum_j \nu_{ij} r_j, \sum_j (-\Delta H_j) r_j$
Flux given by molecular diffusion:	$\nabla J_i, \sum_i J_i \nabla H_i$
Heat transfer by conduction:	$\nabla(\lambda\nabla T)$
Radiation heat flux:	Q_{rad}
Pressure gradient:	∇P

https://doi.org/10.1515/9783110632927-001

Shear stress: ∇s
Gravity: $\rho_f g$

Ideally, with these general equations, it would be impossible to model every kind of reactor. For example, for a batch reactor, the accumulation and the reaction rate are the only terms to be considered, while for a dynamic plug-flow reactor, it is necessary to introduce also the convective term. For multiphase systems, it is necessary to write a mass balance for each phase, including terms related to interfacial mass transfer.

Usually, the reactor models published in the literature are written simplifying the momentum balance equation, often omitted solving the velocity profiles with algebraic expressions (i.e., parabolic profile for laminar flow modeling) and the pressure drop is often calculated by using friction factor correlations [5].

Therefore, in selecting the terms to include in the conservation equations, several aspects must be considered. Some major issues are summarized in Table 1.1.

Table 1.1: Main aspects to be considered in reactor modeling.

	Batch/fed batch	**Continuous reactor**
Solid phase	– Rate laws and temperature dependencies – Adsorption/desorption phenomena – Catalyst deactivation – Intraparticle diffusion limitation – Fluid–solid mass transfer limitation – Change in the physical nature of the fluid–solid film – Formation of a solid shell of another solid compound – Nonuniform active phase – Effective diffusivity determination – Effective thermal conductivity determination – Particle size distribution – Irregular particle shapes	
Fluid phase/s	– Change in the fluid density – Stagnant zones – Hold up – Physical properties determination – Interfacial mass transfer	– Change in the volumetric flow rate – Direction of the fluid – Nonidealities of the flow – Bypasses and stagnant zones – Pressure drop – Hold up – Physical properties determination – Axial and radial dispersion parameters – Interfacial mass transfer

1.2 Why Matlab?

Software devoted to reactor modeling are widely available (MATLAB, gPROMS ModelBuilder, Athena Visual Studio, C++, etc.). The number of physical phenomena to be accounted for in the reactor modeling depends on the user's sensibility. Thus, concentration and temperature gradients could be taken into account when simulating industrial-scale reactors, that is, to avoid a high pressure drop, while it is possible to neglect them when facing with lab-scale reactors. Among the possible choices of modeling tool, Matlab is surely a valid option.

Matlab (**mat**rix **lab**oratory) can be considered as the state-of-the-art software for scientific computation. It is a widely spread software, built as integrated environment where coding, data analysis and visualization can be developed in the same framework. It is constantly developed and updated tool since 80s. Nowadays, it contains a wide collection of already built-in functions tested by Mathworks specialists, allowing the solution for a large number of problems that are normally faced by scientists of each discipline, meaning that the majority of common tasks can be solved with a single function call.

This high flexibility is even empowered by the constantly developed and updated open exchange Matlab Central website, where users share their effort in creating new routines with complete codes and examples of application for many new numerical challenges.

It is used in both academia and industry. It is taught in many university courses both at bachelor's and master's degree levels, starting from the basics and ending with case studies of advanced applications. Despite the fact that the learning curve of this software is relatively fast, many universities split the learning of the software into two- or three-level courses. Usually, at a first level, the main elements of the language are introduced and applied to easy tasks. These elements are then deepened at a second-level course where real case studies are treated, with the necessity to use more sophisticated functions to solve advanced problems.

In addition to the described features, Matlab has the possibility to be easily interfaced with other popular software (e.g., Excel, Comsol, and LabView) enabling the realization of advanced architectures scientific computing.

Many books were already published about the use of Matlab. The majority of them deal with basic teaching and application. A learner who wants to approach Matlab would be spoilt for choice, most of the times ending with the books suggested by Mathworks itself or referring to the well-written Matlab help. On the Mathworks website, it is stated that till January 2020, more than 2,800 books were published dealing with Matlab use for each kind of disciplines, and 51 for chemistry and chemical engineering topics (see Figure 1.1) [6]. Even though this number could be considered high enough to cover the majority of the topics included in chemical engineering area, none of them deal with specific CRE approaches. This discipline is most of the times included in more general books dealing with chemical engineering and/or confined to few pages, including easy examples of application.

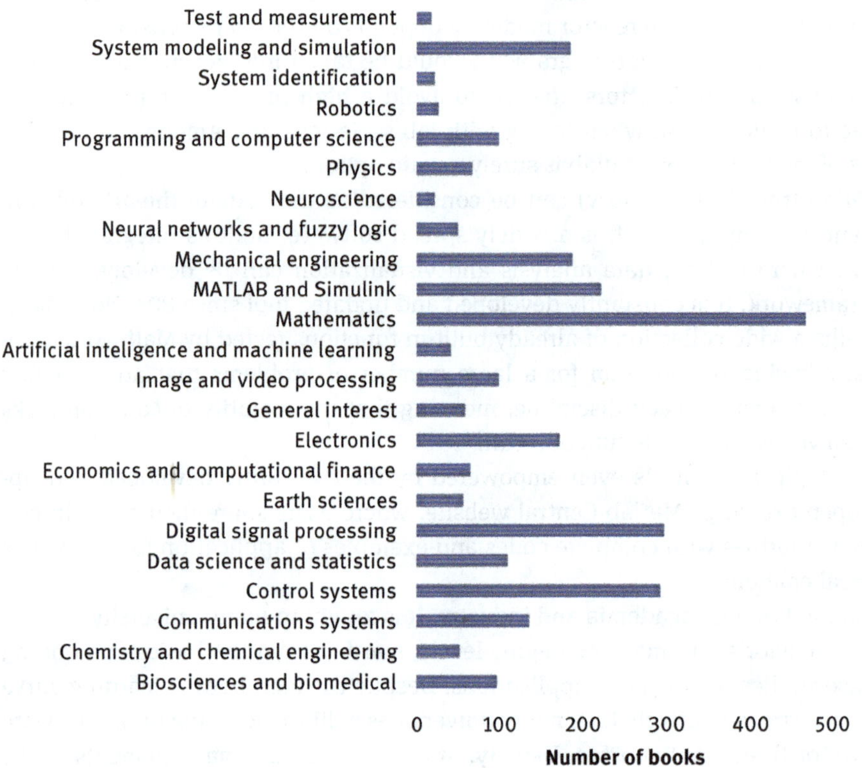

Figure 1.1: Distribution of books dealing with Matlab application in different scientific fields [6].

The main innovation of this book is to extend the treatise to the most common problems in CRE, including advanced approaches to solve the main cases that a specialist of this area could face with. The recipient of this book must be already skilled in coding Matlab; thus, no specific chapter is included in this book. Few hints are given in this textbook about the use of the built-in algorithms. Table 1.2 lists the main Matlab functions used for reactor type.

Table 1.2: Matlab functions used in this book to solve reactor modeling.

Reactor	Task	Matlab function
Batch/fed batch	Simulation	Ordinary differential equation solver (*ode15s*, *ode45*)
	Parameter estimation	Objective function minimization (*lsqnonlin*)
Ideal tubular reactor	Simulation (steady state)	Ordinary differential equation solver (*ode15s*, *ode45*)
	Simulation (dynamic)	Partial differential equation solver (*pdepe*)
	Parameter estimation	Objective function minimization (*lsqnonlin*)
Real tubular reactor	Simulation (steady state)	Partial differential equation solver (*pdepe*)
	Simulation (dynamic)	Method of lines (defined by user)
	Parameter estimation	Objective function minimization (*lsqnonlin*)
Continuous stirred tank reactor	Simulation (steady state)	Nonlinear equation solver (*fzero*, *fsolve*)
	Simulation (dynamic)	Ordinary differential equation solver (*ode15s*, *ode45*)
	Parameter estimation	Objective function minimization (*lsqnonlin*)

The approach adopted in this book is to develop general codes, treating chemical reactions in a schematic simplified way. In general, no real chemical molecules will be considered, as they will be labeled in abstract form,

$$\mathrm{A} + \mathrm{B} \rightarrow \mathrm{C} + \mathrm{D}$$

The advantage of this approach is the realization of general codes, independent from specific physical properties, that allows the writing of smart and compact codes that can be easily adapted by the users for accomplishing their own needs.

List of symbols

C_i	Concentration of component i	[mol/m^3]
$c_{p,i}$	Specific heat of component i	[J/(g K)]
g	Gravitational acceleration constant, 9.822	[m/s^2]
H_i	Partial molar enthalpy	[J/mol]
J_i	Flux given by molecular diffusion	[mol/(m^3 s)]
M_i	Molecular weight of component i	[g/mol]
P	Pressure	[Pa]
Q_{rad}	Radiation heat flux	[J/(m^3 s)]
r_j	Reaction rate	[mol/(m^3 s)]
s	Shear stress	[Pa]
t	Time	[s]
T	Temperature	[K]
u	Fluid velocity	[m/s]

Greek symbols

ΔH_j	Reaction enthalpy of reaction j	[J/mol]
λ	Thermal conductivity	[J/(m s K)]
ρ_f	Fluid density	[kg/m^3]
$\nu_{i,j}$	Stoichiometric coefficient of component i and reaction j	[–]

Mathematical operators

∇ "Nabla" or "del" operator, expressing the gradient of a variable along its coordinates

References

[1] G.F. Froment, K.B. Bischoff, J. De Wilde. Chemical Reactor Analysis and Design (3rd edition). John Wiley & Sons Inc, New York: 2001.

[2] T. Salmi, J.-P. Mikkola, J. Wärnå. Chemical Reaction Engineering and Reactor Technology. CRC Press, New York: 2010.

[3] B. Saha. Catalytic Reactors. Walter de Gruyter, Berlin: 2016.

[4] P.A. Ramachandran, R.V. Chaudari. Three-Phase Catalytic Reactors. Gordon and Breach Science Publishers, New York: 1983.

[5] M. Al-Dahhan, F. Larachi, M. Dudukovic, A. Laurent. High-pressure trickle bed reactors: a review. Industrial and Engineering Chemistry Research 1997, 36, 3292–3314.

[6] https://it.mathworks.com/academia/books (last visited 24.06.2020).

Chapter 2
Batch reactors for homogeneous catalysis

2.1 Introduction

Batch reactor (BR) is a very widely used reactor configuration [1–3], especially for small productions of chemical specialties and industrial syntheses for which different steps and catalysts/additives must be added along the reaction time [4]. Starting from an initial charge of reactants loaded into the reactor, the system evolves with time until the operation is complete, and the products are discharged. Mass and energy balances are then constituted by coupled differential equations to be integrated in time, starting from known initial conditions (initial value problem). In some cases, also of great industrial interest, more than one phase is involved in the operation [5], and the conservation equations for BR must be modified for taking into account for mass transfer from one phase to another. The most common situation involving multiphase reactions are gas–liquid and liquid–liquid. Other exception with respect to the "pure" BR is the fed-batch reactor in which a certain amount of reactant is added (or withdrawn) along the time in form of a continuous feed of pulses.

2.2 Single-phase reactors

2.2.1 Mass balance, isothermal reactor

Consider, as a first approach, a single-phase system that, typically, is in the liquid state. In this case, the mole balance for the ith component can be derived from the general mass conservation equation that can be written as

$$[\text{accumulation}_i] = [\text{inlet}_i] - [\text{outlet}_i] + [\text{generation}_i] \tag{2.1}$$

A similar equation can be written for the total mass instead of that for a single component, even if it is less useful. Equation (2.1) is a general form for the mass balance of a chemical substance, and in common reactor configurations, some terms can be eliminated. It is evident that for a BR, for which no inlet and outlet terms are present, the previous equation reduces to:

$$[\text{accumulation}_i] = [\text{generation}_i] \tag{2.2}$$

In an explicit form, eq. (2.2) is

$$\frac{dn_i}{dt} = V_R \nu_i r \tag{2.3}$$

https://doi.org/10.1515/9783110632927-002

where n_i is the mole number of ith component, t is the time, V_R is the reaction volume, v_i is the stoichiometric coefficient of component i in the single occurring reaction while r is the reaction rate. For a reactive system in which Nc component is simultaneously present, Nc-coupled ordinary differential equation (ODEs) like eq. (2.3) must be solved by integration into time. In the derivation of eq. (2.3), the assumption of a perfect mixing and spatial homogeneity was assumed. Some generalization can be further introduced in relation (2.3) by considering the following points:

- Equation (2.3) is valid for a constant volume system, for example, when the density difference between the reactants and the products is low or when the reaction involves a solvent that is therefore the most abundant component. When these assumptions are not valid, reaction volume variation must be taken into account. This can be made by updating the volume V_R during the time by adopting the approximation of volume additivity, like the following:

$$V_R = \sum_{j=1}^{Nc} V_j = \sum_{j=1}^{Nc} \frac{n_j M w_j}{\rho_j} \tag{2.4}$$

  where Mw_j is the molecular weight of component j and ρ_j is its density. During the integration of ODE system (2.3), eq. (2.4) can be used for updating reaction volume at each integration step in time. Obviously, if the system is not perfectly isothermal (see Section 2.2.2) densities as functions of temperature can be introduced in eq. (2.4) for a correct representation of volume variation. It is obvious that in this kind of situation, the *reaction* volume is different from reactor volume.
- A second generalization regards the possibility to account for the occurrence of several chemical reactions. In this case, the generation term by chemical reactions can be written as follows:

$$R_i = \sum_{k=1}^{Nr} v_{i,k} r_k \tag{2.5}$$

  where Nr is the number of independent chemical reactions, R_i is the overall generation/consumption term for component i, r_k is the kth reaction rate and $v_{i,k}$ is the stoichiometric coefficient of component i in the reaction k (frequently referred to as stoichiometric matrix).

From a practical point of view, the integration of the ODE system (2.3), both in the case of single and multiple reactions, requires the knowledge of the kinetic expression for r_i and the related kinetic parameters. The most common form for a simple kinetic expression of a chemical reaction is represented by the following equation:

$$r_i = k_i \prod_{j=1}^{Nc_i} C_j^{\lambda_j} \tag{2.6}$$

where k_i is the kinetic constant, Nc_i is the number of components involved in the reaction i, C_j is the concentration of the components and λ_j is reaction order with respect to the component i. Obviously, instead of concentrations, in eq. (2.6), mole numbers can be directly substituted instead of concentrations giving place to the following expression:

$$r_i = k_i \prod_{j=1}^{Nc_i} \left(\frac{n_j}{V_R}\right)^{\lambda_j} \tag{2.7}$$

As stated earlier, for a variable volume system, the reaction volume can be calculated by eq. (2.4). When the system of ODEs describing the BR is numerically solved (eq. (2.3)), the time profile for the concentrations, or mole number, of all components is available. From these profiles, it is straightforward to obtain profiles for conversion, selectivity, yields and other derived quantities.

The system considered in the exposed treatment consists in a single or multiple reaction occurring in a homogeneous liquid phase, that is, the more frequent situation. However, if the chemical reaction occurs in a gas phase, the mass balance equations (2.3) remain valid and the only modification is related to the kinetic expression that is expressed, very likely, in terms of partial pressures as follows:

$$r_i = k_i \prod_{j=1}^{Nc_i} p_j^{\lambda_j} = k_i \prod_{j=1}^{Nc_i} (Py_j)^{\lambda_j} \tag{2.8}$$

where p_j is the partial pressure of jth component, P is the total pressure and y_j is the mole fraction in gaseous phase related to component j. For BR operating in gaseous phase, the reaction volume corresponds to the whole reactor volume but, according to the reaction stoichiometry, the total pressure could vary.

Example 2.1 Single reaction, variable volume, liquid phase, isothermal reactor

In a BR, a single reaction occurs with the following stoichiometry:

$$A + B \rightarrow C$$

This reaction is characterized by a second-order kinetics with a reaction rate expression expressed by the relation ($k = 1e{-}6\ m^3/(mol\ s)$):

$$r = kC_A C_B$$

The properties of the three components and the initial charge in the reactor are summarized in Table 2.1:

Table 2.1: Initial charge in the batch reactor, Example 2.1.

Component	Initial moles (mol)	Molecular weight (kg/mol)	Density (kg/m^3)
A	1,000	0.028	570
B	1,200	0.018	1,000
C	0	0.046	789

To describe the time profiles of both components' concentration and liquid volume, the following equation must be solved:

$$\frac{dn_A}{dt} = -V_R r \quad \frac{dn_B}{dt} = -V_R r \quad \frac{dn_C}{dt} = +V_R r$$

$$V_R = \sum_{i=1}^{N_C} \frac{n_i M_{Wi}}{\rho_i} \quad C_i = \frac{n_i}{V_R}$$

Matlab code for the solution of this example and results are reported as follows and results (Figure 2.1):

```
% example 2.1
clc,clear
global mw ro ni k
mw=[0.028 0.018 0.046]; % molecular weights (kg/mol)
ro=[570 1000 789];      % densities (kg/m3)
ni=[-1  -1   1];        % stoich. Coefficients (-)
n0=[1000 1200 0];       % initial moles (mol)
tspan=0:0.1:800;        % time span (s)
k=1e-6;                 % kinetic constant (m3/(mol s))
[tx,nx]=ode45(@ode_ex2_01,tspan,n0);
for j=1:length(nx)
  Vr(j)=0;
  for k=1:3
   Vr(j)=Vr(j)+nx(j,k)*mw(k)/ro(k);
  end
end

%% plot
subplot(1,2,1)
plot(tx,nx(:,1),tx,nx(:,2),tx,nx(:,3))
grid
xlabel('Time (s)')
ylabel('Moles (mol)')
```

```
legend('A','B','C')
subplot(1,2,2)
plot(tx,Vr)
grid
xlabel('Time (s)')
ylabel('Reaction volume (m3)')

function [dn] = ode_ex2_01(t,n)
global mw ro ni k
Vr=sum(n'.*mw./ro);
c=n/Vr;
r=k*c(1)*c(2);
dn(1)=ni(1)*Vr*r;
dn(2)=ni(2)*Vr*r;
dn(3)=ni(3)*Vr*r;
dn=dn';
end
```

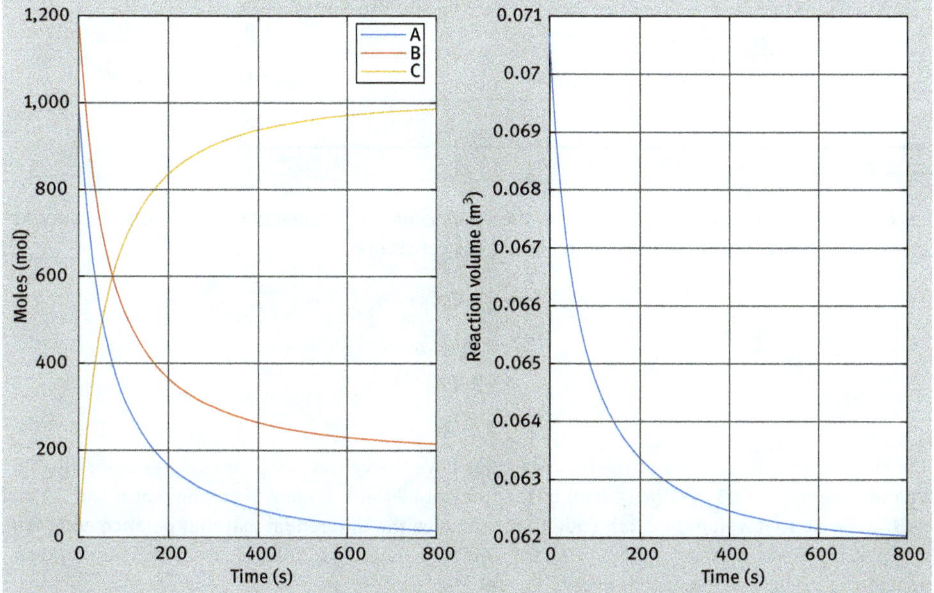

Figure 2.1: (Left) Calculated profiles of the amount of substance in moles of each component versus time; (right) reaction volume versus time.

Example 2.2 Multiple reactions, constant volume, liquid phase, isothermal reactor
In a BR, a set of four multiple reactions in a complex scheme occur involving five components identified as A, B, C, D and E. The reactions scheme is as follows:

$$A + B \longleftrightarrow 2C + D$$
$$A + C \longrightarrow E$$
$$D \longrightarrow 2B$$

The first reaction is reversible, so forward and reverse reactions must be accounted for. On the basis of this scheme, the following stoichiometric matrix can be arranged in which the stoichiometric coefficients of all components in all reactions are included (Table 2.2).

Table 2.2: Stoichiometric matrix for the reaction system of Example 2.2.

Component	Reaction 1	Reaction 2	Reaction 3	Reaction 4
A	−1	+1	−1	0
B	−1	+1	0	+2
C	2	−2	−1	0
D	1	−1	0	−1
E	0	0	1	0

For a complete definition of the problem, the expressions of the reaction rates are necessary, together with the values of the corresponding kinetic constants:

$$r_1 = k_1 C_A C_B$$
$$r_2 = k_2 C_C^2 C_D$$
$$r_3 = k_3 C_A$$
$$r_4 = k_4 C_D$$

The kinetic constants related to the four reactions have the values: 1*e*–5, 2*e*–6 3*e*–5 1*e*–5; the reaction volume is of 0.003 m^3. The reactor is charged initially with 1.5 mol of component A and 1.2 mol of B. We wish to develop a Matlab code that can solve the differential material balance equations and plot the concentration profiles of all the components for a 1,200 s of reaction time. In a compact form, the material balance for BR is

$$\frac{dn_i}{dt} = V_R \sum_{k=1}^{Nr} v_{i,k} r_k, \quad i = \text{A, B, C, D, E}, \quad N_R = \text{four reactions}$$

Matlab code for the solution of this example and results are reported as follows:

```
% example 2.2
clc,clear
global k Vr ni Nc Nr
Nc=5;                              % number of components (-)
Nr=4;                              % number of reactions (-)
Vr=0.003;                          % reaction volume (m3)
k=[0.01 0.002 0.03 0.01]/1000; % kinetic constants
ni=[-1 +1 -1 0;                    % matrix of stoichiometric coefficients
   -1 +1 0 +2;                     % row  -> component
   +2 -2 -1 0;                     % column -> reaction
   +1 -1 0 -1
    0 0 +1 0];

n0=[1.5 1.2 0 0 0];  % initial moles (mol)
tspan=0:0.1:1200;  % time range for integration (s)
[tx,nx]=ode15s(@ode_ex2_02,tspan,n0);
%% plot
subplot(1,2,1)
plot(tx,nx(:,1),tx,nx(:,2),tx,nx(:,3))
grid
xlabel('Time (s)')
ylabel('Moles (mol)')
legend('A','B','C')

subplot(1,2,2)
plot(tx,nx(:,4),tx,nx(:,5))
grid
xlabel('Time (s)')
ylabel('Moles (mol)')
legend('D','E')

function [dn] = ode_ex2_02(t,n)
%% ---------------------------------------
% example 2.2 - equations
%----------------------------------------
global k Vr ni Nc Nr
c=n/Vr;
r(1)=k(1)*c(1)*c(2);
r(2)=k(2)*c(3)^2*c(4);
r(3)=k(3)*c(1)*c(3);
r(4)=k(4)*c(4);
for j=1:Nc
  dn(j)=0;
  for jj=1:Nr
    dn(j)=dn(j)+Vr*r(jj)*ni(j,jj);
  end
end
```

```
dn=dn';
end
```

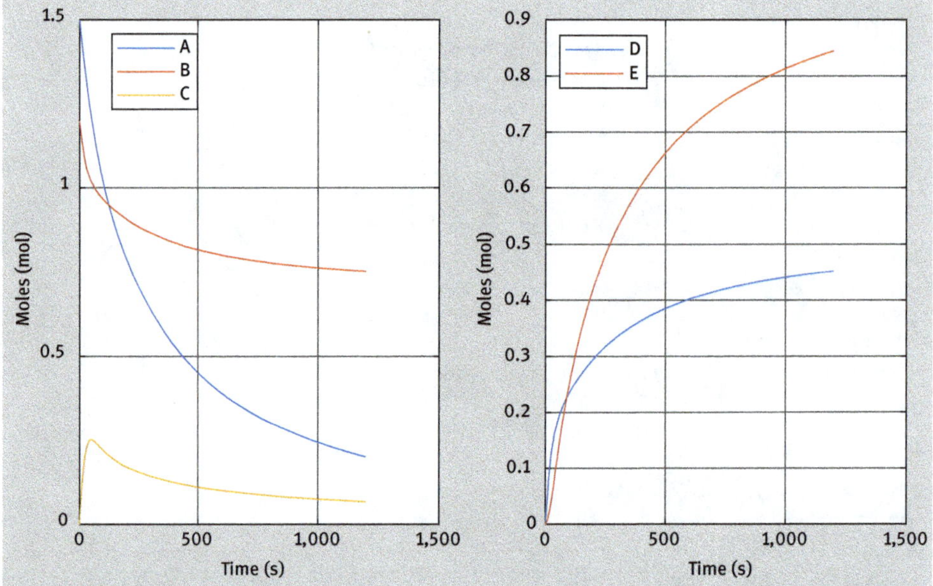

Figure 2.2: (Left) Number of moles of profiles for components A, B and C versus time; (right) number of moles of profiles for components D and E versus time.

Example 2.3 Single reaction, variable pressure, gas phase, isothermal reactor

In a BR, a single gaseous phase reaction occurs with the following stoichiometry:

$$A + B \rightarrow C + 2D$$

This reaction is characterized by a second-order kinetics with a reaction rate expression represented by the following relation in which the kinetic constant is $k = 0.5$ mol/(m^3 s atm^2):

$$r = kP_A P_B$$

The moles of each component initially charge in the reactor are summarized in Table 2.3:

Table 2.3: Initial charge in the batch reactor, Example 2.3.

Component	Initial moles (mol)
A	100
B	120
C	0
D	0

Being the reaction that occurs in gaseous phase, the reaction volume is equivalent to the whole reactor volume. To describe the time profiles of both components' concentration and total pressure, the following coupled differential equations must be solved:

$$\frac{dn_A}{dt} = -V_R r \quad \frac{dn_B}{dt} = -V_R r \quad \frac{dn_C}{dt} = +V_R r \quad \frac{dn_D}{dt} = +V_R 2r$$

$$P_{tot} = \frac{R_G T}{V_R}\left(\sum_{i=1}^{N_C} n_i\right) \quad y_i = \frac{n_i}{\sum_{i=1}^{N_C} n_i} = \frac{n_i}{n_{tot}} \quad p_i = P_{tot} y_i \quad i = \text{A, B, C, D}$$

Matlab code for the solution of this example and results (Figure 2.3) are reported as follows:

```
% example 2.3
clc,clear
global Vr T ni Rg k
Vr=3;                % reactor volume (m3)
ni=[-1 -1 1 2];      % stoich. coefficients (-)
Rg=0.08205/1000;     % gas constant (atm m3/(mol K))
T=340;               % temperature (K)
k=0.5;               % kinetic constant (mol/(s m3 atm2)
n0=[100 120 0 0];    % initial moles (mol)
tspan=0:0.1:800;     % time range for integration (s)
[tx,nx]=ode45(@ode_ex2_03,tspan,n0);
for j=1:length(nx)
  ntot(j)=sum(nx(j,:));        % total moles (mol)
  Ptot(j)=ntot(j)*Rg*T/Vr;     % total pressure (atm)
end

%% plot
subplot(1,2,1)
plot(tx,nx(:,1),tx,nx(:,2),tx,nx(:,3),tx,nx(:,4))
grid
xlabel('Time (s)')
ylabel('Moles (mol)')
```

```
legend('A','B','C','D')
subplot(1,2,2)
plot(tx,Ptot)
grid
xlabel('Time (s)')
ylabel('Total pressure (atm)')

function [dn] = ode_ex2_03(t,n)
global Vr T ni Rg k

ntot=sum(n);       % total moles (mol)
y=n/ntot;          % mole fraction (-)
P=ntot*Rg*T/Vr;    % total pressure (atm)
p=P*y;             % component partial pressure (atm)

r=k*p(1)*p(2);     % reaction rate (mol/(m3 s))
dn=ni*Vr*r;        % mass balance (mol/s)
dn=dn';
end
```

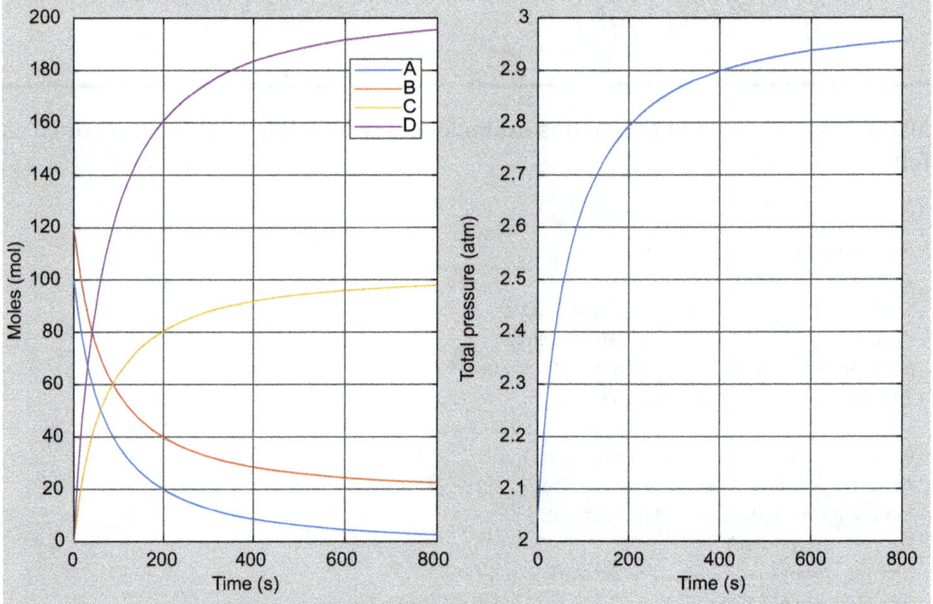

Figure 2.3: (Left) Number of moles profiles for all the components versus time; (right) total pressure versus time.

2.2.2 Energy balance

In Section 2.2.1, the BR has been assumed as isothermal and the attention was only focused on mass conservation. According to the mathematical treatment already presented, the fixed temperature is expressed by the fixed kinetic constant k that appears in eqs. (2.6) and (2.7). When we move from isothermal to nonisothermal reactor, we must couple to the mass balance and energy balance conservation equation [1]. In the present section, we will derive such relation that, depending on the heat exchange characteristics, is able to describe thermal behavior of the BR reactor configuration.

Starting from a general heat balance equation such as:

$$[\text{heat accumulated}] = [\text{inlet heat}] - [\text{outlet heat}] + [\text{heat generated}] \tag{2.9}$$

Due to the intrinsic setup of the BR, inlet and outlet heat are terms that can be assumed null so the eq. (2.9) reduces to

$$[\text{heat accumulated}] = [\text{heat generated}] \tag{2.10}$$

In explicit form, the terms of eq. (2.10) can be expressed as

$$mC_{\text{pm}}\frac{dT}{dt} = r(-\Delta H_r)V_R - Q_E \tag{2.11}$$

where m is the mass of the reaction mixture in the reactor, C_{pm} is the average specific heat of the reacting mixture, T is the temperature, ΔH_r is the reaction heat and Q_E is the heat exchanged between the reactor and the surrounding environment. Equation (2.11) contains the assumption that the system is spatially homogeneous from the point of view of temperature, so the temperature is assumed equal from point to point in the reactor. The thermal behavior of the reactor is contained in the expression of the Q_E term for heat exchanged:

$$Q_E = UA(T - T_S) \tag{2.12}$$

where U is the overall heat exchange coefficient, A is the exchange area and T_S is the temperature of the surrounding environment. It should be noted that the temperature of the surrounding environment (for example the temperature of a heating medium flowing in the jacket) could itself be a function of time. This aspect represents a further source of variation for the exchanged heat Q_E. According to the difference between T and T_S, the heat is transferred from the external environment to the reactor or inversely. The value of the overall heat transfer coefficient U can represent the discrimination between isothermal, adiabatic or intermediate behavior as in the following schematization:

- $U = 0$ adiabatic reactor, no heat is transferred from/to the reactor
- $U = +\infty$ isothermal reactor, all the heat generated is removed by the thermal control system

- U = value of the reactor shows an intermediate behavior between isothermal and adiabatic one

For a practical purpose and in order to integrate eq. (2.11), temperature derivative must be explicitly written and eq. (2.12) must also been introduced in the operative heat balance equation:

$$\frac{dT}{dt} = \frac{[r(-\Delta H_r)V_R - UA(T - T_S)]}{mC_{\text{pm}}} \tag{2.13}$$

When this heat balance equation is applied to a system with variable volume, not only V_R must be evaluated along the time with the relation (2.4), but also the exchange area A could be adjusted according to the volume in the reactor and the related liquid level. This last should be a relatively small effect but in a rigorous model should be considered. Another possible refinement of the model consists in using, for the specific heat of the reacting mixture, a function of temperature and composition instead of an average value. Total mass in the reactor can be eliminated from eq. (2.13) by using the following relation that introduce the mixture density ρ_m:

$$\rho_m = \frac{m}{V_R} \tag{2.14}$$

As done for specific heat, also mixture density can be expressed as a function of temperature and composition in order to obtain a more rigorous description of the system. By substituting eq. (2.14) into (2.13), we obtain:

$$\frac{dT}{dt} = \frac{1}{\rho_m C_{\text{pm}}}\left[r(-\Delta H_r) - \frac{UA}{V_R}(T - T_S)\right] \tag{2.15}$$

Equation (2.15) is valid only for a single chemical reaction that occurs inside the BR vessel. In the more general case when multiple reactions must be considered, equation (2.15) must be slightly modified by accounting for all the thermal effects connected to the reactions network:

$$\frac{dT}{dt} = \frac{1}{\rho_m C_{\text{pm}}}\left[\sum_{j=1}^{N_r} r_j\left(-\Delta H_{r_j}\right) - \frac{UA}{V_R}(T - T_S)\right] \tag{2.16}$$

The temperature T_S in eq. (2.16) can fall into one of the three different possibilities:

- Fixed value
- Value that is a function of time according to a ramp or steps
- Derived from an energy balance related to heating jacket. In this last case, a differential equation similar to (2.16), but without reaction term, must be coupled to the already defined mass and energy balance ODEs.

Equation (2.15) (or alternatively (2.16)) must be coupled with the mass balance for each component (2.3) obtaining in such a way a system of $Nc+1$ ODEs that can be solved numerically to calculate simultaneously time profiles for component moles and system temperature.

It is worth mentioning that the system temperature affects in a strong way also the reaction rates through the temperature dependence of the kinetic constant. The well-known Arrhenius equation expresses this dependency:

$$k_i = k_{iref} \exp\left[\frac{E_{Ai}}{R}\left(\frac{1}{T_{ref}} - \frac{1}{T}\right)\right] \tag{2.17}$$

Where k_{iref} is the kinetic constant referred to a reference temperature T_{ref}, R is the gas constant, T is the absolute temperature and E_{Ai} is the activation energy for the reaction i.

Example 2.4 Multiple reactions, constant volume, liquid phase, nonisothermal reactor
In a constant volume BR, two coupled chemical reactions occur with the following stoichiometry:

$$A + B \rightarrow C$$
$$A + 2C \rightarrow D + E$$

These reactions are characterized by a kinetics reported in Table 2.4, in terms of reaction rate expression and kinetic parameters:

Table 2.4: Kinetic expression and parameters, Example 2.4.

Reaction	Rate expression	k_{ref}	*Ea* (J/mol)	*ΔHr* (J/mol)
1	$r = kC_AC_B$	5e–5 ($m^3/(mol\ s)$)	42,500	–85,000
2	$r = kC_AC_C^2$	2e–6 ($m^6/(mol^2\ s)$)	44,500	–80,000

The initial charge in the reactor is reported summarized in Table 2.5:

Table 2.5: Initial charge in the batch reactor of Example 2.4.

Component	Initial moles (mol)
A	1,000
B	900
C	150
D	20
E	0

Other useful data regarding reactor characteristics and physicochemical properties of the mixture are summarized in Table 2.6.

Table 2.6: Physical properties and reactor settings, Example 2.4.

Property	Value	Units
Reactor volume	2	(m^3)
Overall heat transfer coefficient	5,000	(J/(s m^2 K))
Heat exchange area	20	(m^2)
Average density of mixture	1,100	(kg/m^3)
Average specific heat	4,000	(J/(kg K))
Heating/cooling fluid temperature	340	(K)
Initial reactor temperature	330	(K)
Time span for integration	0–800	(s)

To describe the time profiles of both components' concentration and reaction temperature, the following equation must be solved simultaneously:

$$\frac{dn_A}{dt} = V_R(-r_1 - r_2) \quad \frac{dn_B}{dt} = V_R(-r_1) \quad \frac{dn_C}{dt} = V_R(+r_1 - 2r_2) \quad \frac{dn_D}{dt} = V_R(+r_2) \quad \frac{dn_E}{dt} = V_R(+r_2)$$

$$\frac{dT}{dt} = \frac{Q_r - Q_E}{\rho_m C_{pm}}$$

$$C_i = \frac{n_i}{V_R} \quad Q_E = \frac{UA}{V_R}(T - T_S) \quad Q_r = -r_1 \Delta H_{r1} - r_2 \Delta H_{r2}$$

Matlab code for the solution of this example and results (Figure 2.4) are reported as follows:

```
% example 2.4
clc,clear
global Vr U DH A Ts rom cpm Ea Tref kref

Vr=2;                  % reaction volume (m3)
U=5000;                % overall heat transfer coefficient (J/(s m2 K))
DH=[-85000 -80000];    % heat of reactions (J/mol)
rom=1100;              % average mixture density (kg/m3)
cpm=4000;              % average liquid heat capacity (J/(kg K))
A=20;                  % heat transfer area (m2)
Ts=340;                % temperature of jacket fluid (K)
Ea=[42500 44500];      % activation energies (J/mol)
Tref=323;              % reference temperature (K)
```

```
kref=[5e-5 2e-6];

n0=[1000 900 150 20 0]; % initial moles (mol)
t0=330;                 % initial temperature (K)
y0=[n0 t0];             % initial conditions
tspan=0:0.01:400;       % time interval for integration (s)
[t,y]=ode15s(@ode_ex2_04,tspan,y0);

%% plots
subplot(1,2,1)
plot(t,y(:,1:5))
grid
xlabel('Time (s)')
ylabel('Moles (mol)')
legend('A','B','C','D','E')

subplot(1,2,2)
plot(t,y(:,6))
grid
xlabel('Time (s)')
ylabel('Temperature (K)')

function [dy] = ode_ex2_04(t,y)
global Vr U DH A Ts rom cpm Ea Tref kref

n=y(1:5);       % moles
T=y(6);         % temperature
c=n/Vr;         % concentrations
R=1.987*4.189; % gas constant

k(1)=kref(1)*exp((Ea(1)/R)*(1/Tref-1/T));
k(2)=kref(2)*exp((Ea(2)/R)*(1/Tref-1/T));
r1=k(1)*c(1)*c(2);
r2=k(2)*c(1)*c(3)^2;
dy(1)=Vr*(-r1-r2);
dy(2)=Vr*(-r1);
dy(3)=Vr*(+r1-2*r2);
dy(4)=Vr*(+r2);
dy(5)=Vr*(+r2);
Qs=U*A/Vr*(T-Ts);
Qr=-r1*DH(1)-r2*DH(2);

dy(6)=(Qr-Qs)/(rom*cpm);
```

```
dy=dy';
end
```

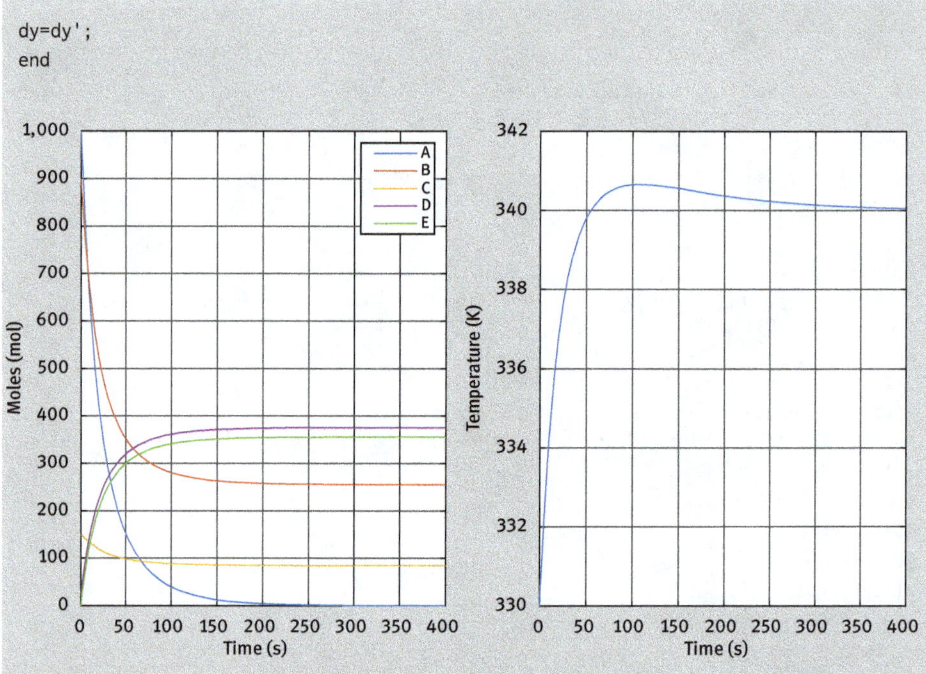

Figure 2.4: (Left) Number of moles profiles of all the components versus time; (right) reaction temperature versus time.

2.3 Two-phase reactors

In a BR, a particular situation can occur when more than a single phase is present. In this chapter, the simultaneous presence of gas and liquid or two liquid phases will be considered while the presence of a solid phase (usually the catalyst for the reaction) will be the subject of successive chapters on reactors for heterogeneous catalysis (Chapters 3 and 6). For gas–liquid system, we assume that the chemical reactions take place only in the liquid phase. On the contrary, for liquid–liquid system, the reactions can, in principle, occur in both liquid phases.

2.3.1 Gas–liquid two-phase reactor

From a general point of view, when two fluid phases are simultaneously present in a BR, different theories can be adopted for the description of mass transfer across the interface [6] and the most widely diffused is the Withman two-film theory [7,8].

According to this theory, in stationary state, two stagnant films are considered on both sides of the separation interface, one for gas and another for liquid phase. In each of these films, the mass transport occurs only by molecular diffusion, and linear pressure and concentration gradients are developed, respectively for the gas and the liquid. Apart from these films, the bulk of the two phases are considered perfectly mixed and no gradients are assumed from point to point in the bulk. A scheme of the described assumption is reported in the following Figure 2.5, referred to a transport of a generic component i in direction x (gas to liquid).

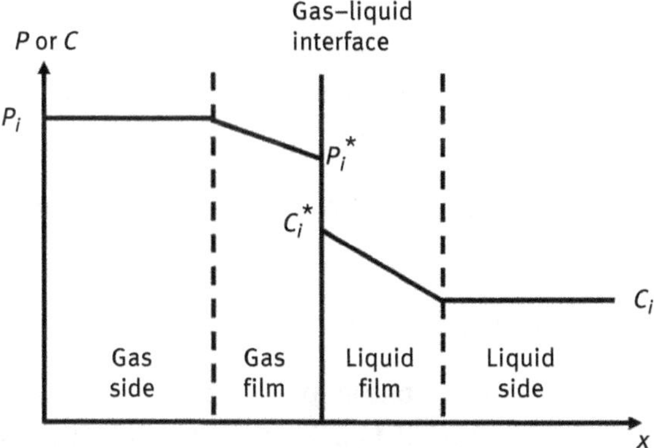

Figure 2.5: Scheme for mass transport in gas–liquid system. Both gas-side and liquid-side films are considered.

It is obvious that the mass transport occurs in the direction of the gradient, that is, from more concentrated to less concentrated portion of the reactor. Moreover, very frequently, the mass transfer resistance from gas side is negligible [9–11] and the overall resistance in located in the liquid side. In this last case, the scheme representing mass transport must be modified and is reported in Figure 2.6.

In Figures 2.5 and 2.6, P_i is the partial pressure of i, P_i^* is the equilibrium pressure, C_i^* is the saturation concentration and C_i is the actual concentration. When two phases are present, it is important to express the transferred quantity of each component to correctly introduce it in the material balance equation. According to the scheme in Figure 2.6, the amount of i transferred per unit of time and per unit of surface (orthogonal to x direction of transport) can be expressed as follows:

$$J_i = k_{Li}a(C_i^* - C_i) \tag{2.18}$$

In this case, only one transport step must be considered, that is liquid side, as the gas-side resistance has been neglected. In a more general situation also, gas-side coefficient should be used. In eq. (2.18) J_i is the transport flux of component i, k_{Li} is the gas–liquid (liquid side) mass transfer coefficient and a is the specific mass transfer area. Many common cases could be useful to lump mass transfer coefficient

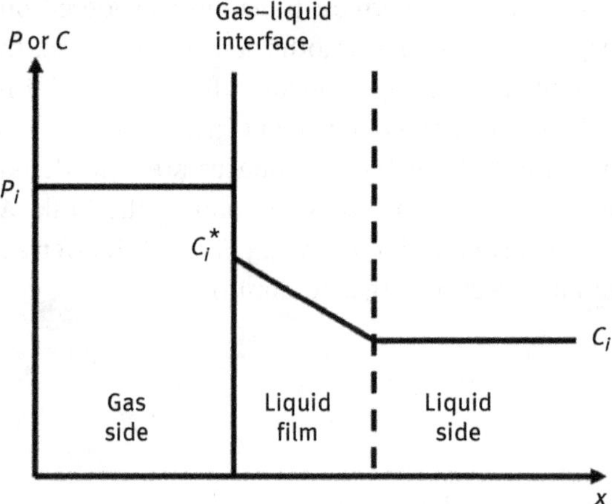

Figure 2.6: Scheme for mass transport in gas–liquid system. Mass transfer resistance is only in liquid side.

k_{Li} and specific area in a single volumetric transport coefficient, β, and the resulting equation is then

$$J_i = \beta_i (C_i^* - C_i) \tag{2.19}$$

Equation (2.19) could be more practical to introduce partial pressure of component i and solubility constant (Henry's constant) instead of C_i^*. Equation (2.19) can then be rewritten as

$$J_i = \beta_i \left(\frac{P_i}{H_i} - C_i \right) \tag{2.20}$$

In eq. (2.20), the Henry's constant H for various components can be assumed effectively a constant or can be expressed as a function of temperature and liquid composition for an accurate description of the solubility. The phenomenon of mass transfer is often coupled with chemical reaction that occurs in the liquid phase containing some reactants, while other reactants are initially present in gas phase. Then, gas–liquid mass transfer has the role to supply reactant for the chemical reaction occurring in the liquid phase. For modeling purposes, two types of situations will be considered in this chapter, as depicted in Figure 2.7(a) and (b).

In Figure 2.7(a), a constant pressure operation is schematized. The gaseous reactant is fed to the reactor through a control valve driven, in a control loop, by a pressure transducer in order to maintain the pressure constant through the entire operation. The feed is necessary as the mass transfer/consumption of reactant tend to deplete the gaseous atmosphere. On the contrary, in Figure 2.7(b), no gaseous reactant is fed to the reactor and the pressure in the gas phase will decrease along the time.

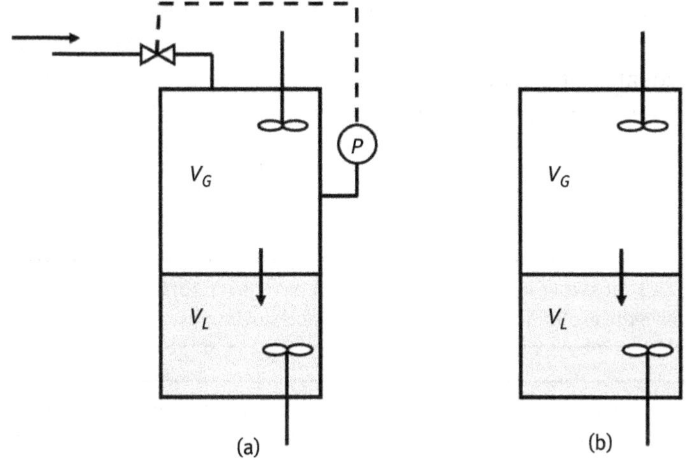

Figure 2.7: (a) Constant pressure operation and (b) variable pressure operation.

In the case of constant pressure (scheme (a) in Figure 2.7), the mass balance on gas phase is not necessary, while that on the liquid phase equation (2.3) must be modified as follows:

$$\begin{cases} \dfrac{dn_i}{dt} = V_L v_i r & \text{for components not affected by mass transfer} \\ \dfrac{dn_i}{dt} = V_L v_i r + J_i V_L & \text{for components affected by mass transfer} \end{cases} \tag{2.21}$$

where J_i can be evaluated through relation (2.20). In the derivation of eq. (2.21), the assumption of constant volumes for both phases was introduced.

When the reactor is operated at variable pressure, without the controlled feed of gaseous reactant, the gaseous phase is depleted, and the pressure decreases as the operation proceed. This aspect is particularly important, for modeling purposes, as the pressure in contained into eq. (2.20) and a decrease of pressure involve a decrease in mass-transfer driving force that must be modeled accordingly. A balance on gas phase must be coupled with other balance equations, already derived for liquid phase, and the following relation can be written as

$$\frac{dn_i^G}{dt} = -J_i V_L \tag{2.22}$$

The integration of eq. (2.22) allows for the calculation of moles in gas phase from which pressure can be calculated and used in eq. (2.20):

$$P_i = \frac{n_i^G RT}{V_G} \tag{2.23}$$

Equation (2.23) involves the assumption of an ideal behavior of the gaseous phase even if more complex and accurate equations of state can be used for the description of the eventual nonideality of gas mixture. As an alternative to eq. (2.23), a further differential equation can be coupled to the already developed, that is:

$$\frac{dP_i}{dt} = \frac{RT}{V_G}\frac{dn_i^G}{dt} \tag{2.24}$$

The system of ODEs (2.21)–(2.23), together with the auxiliary algebraic eqs. (2.20) and (2.23), must be solved simultaneously to give the evolution in time of all the variables of interest.

Example 2.5 Gas–liquid single reaction, constant volume, variable pressure, isothermal reactor
A gas–liquid isothermal reactor is operated with a variable pressure in vapor phase with a single chemical reaction described by the following stoichiometry:

$$A + B \rightarrow C$$

Reactant B is initially charged in the reactor and, at time = 0, is present only in gas phase. This reactant is gradually transferred to the liquid phase in which the reaction occurs. The kinetic of the reaction is represented by a second-order expression with a constant as in Table 2.7:

Table 2.7: Rate equation and kinetic parameter for Example 2.5.

Reaction	Rate expression	k
1	$r = kC_AC_B$	1.2e−5 ($m^3/(mol\ s)$)

The initial charge in the reactor (liquid phase) is summarized in Table 2.8:

Table 2.8: Initial charge in the batch reactor, Example 2.5.

Component	Initial moles (mol)
A	200
B	0
C	0

Other useful data regarding reactor characteristics and physicochemical properties of the mixture are summarized in Table 2.9.

Table 2.9: Physical properties and reactor settings, Example 2.5.

Property	Value	Units
Reactor volume	2	(m^3)
Liquid volume	1.2	(m^3)
Reactor temperature	360	(K)
Initial pressure of B in gas phase	3	(atm)
Henry's constant	20	(mol/(m^3 atm))
Time span for integration	0–1,800	(s)
Volumetric mass transfer coefficient	0.01	(s^{-1})

Part 1

In a first subplot, describe the time profiles of the moles of three components (in liquid phase) in the time range reported in the previous table. In a second subplot report, the profile of moles of B in gas phase.

For solving the problem in part 1 and assuming as negligible the gas-side mass transfer resistance, the following equations must be solved simultaneously:

$$\frac{dn_A^L}{dt} = V_L(-r) \quad \frac{dn_B^L}{dt} = V_L(-r) + V_L J_B \quad \frac{dn_C^L}{dt} = V_R(+r) \quad \frac{dn_B^G}{dt} = -V_L J_B$$

$$J_B = \beta(C_B^* - C_B) \quad C_B^* = K_H P_B \quad P_B = \frac{n_B^G RT}{V_G} \quad V_G = V_R - V_L$$

Part 2

Perform a parametric study by solving the same problem defined in part 1 but with various values assigned to β as in the subsequent list: 0.1; 0.01; 0.001; 0.0005; 0.0001. Report in a third subplot, the trend of moles of component A in liquid phase for the various values of mass transfer coefficients.

Matlab code for the solution of this example and results (Figure 2.8) are reported as follows:

```
% example 2.5
clc,clear

global k Vr Vl beta T Kh Rgas

%% part 1
Vr=2;          % overall reactor volume (m3)
Vl=1.2;        % liquid phase volume (m3)
k=1.2e-5;      % kinetic constant (m3/(mol s))
beta=0.01;     % G-L mass transfer coefficient (s)
T=360;         % temperature (K)
Kh=20;         % Henry's constant (mol/(m3 atm))
pB0=3;         % initial pressure of B (atm)
```

```
Rgas=0.08205/1000; % gas constant (atm m3/(mol K))
Vv=Vr-Vl;           % gas phase volume (m3)

nAl0=200;               % initial moles of A in liquid (mol)
nBl0=0;                 % initial moles of B in liquid (mol)
nCl0=0;                 % initial moles of C in liquid (mol)
nBv0=pB0*Vv/(Rgas*T); % initial moles of B in gas (mol)

n0=[nAl0 nBl0 nCl0 nBv0];
tspan=0:0.1:1800;

[tx,nx]=ode15s(@ode_ex2_05,tspan,n0);

%% part 2 - parametric study of beta
betax=[0.1 0.01 0.001 0.0005 0.0001];
for j=1:length(betax)
  beta=betax(j);
  [tx,nxx]=ode15s(@ode_ex2_05,tspan,n0);
  na(:,j)=nxx(:,1);
end

%% plot
subplot(1,3,1)
plot(tx,nx(:,1),tx,nx(:,2),tx,nx(:,3))
grid
xlabel('Time (s)')
ylabel('Moles in liquid phase (mol)')
legend('Al','Bl','Cl')

subplot(1,3,2)
plot(tx,nx(:,4))
grid
xlabel('Time (s)')
ylabel('Moles of B in vapor phase (mol)')
legend('Bv')

subplot(1,3,3)
plot(tx,na)
grid
xlabel('Time (s)')
ylabel('Moles of A in liquid phase (mol)')
legend('beta=0.1','beta=0.01','beta=0.001','beta=0.0005','beta=0.0001')

function [dn] = ode_ex2_05(t,n)
global k Vr Vl beta T Kh Rgas

nAl=n(1);
nBl=n(2);
```

```
nCl=n(3);
nBv=n(4);
Vv=Vr-Vl;
cA=nAl/Vl;
cB=nBl/Vl;
cC=nCl/Vl;
pB=nBv*Rgas*T/Vv;
r=k*cA*cB;
cBs=Kh*pB;
J=beta*(cBs-cB);
dn(1)=-Vl*r;
dn(2)=-Vl*r+Vl*J;
dn(3)=+Vl*r;
dn(4)=-Vl*J;
dn=dn';
end
```

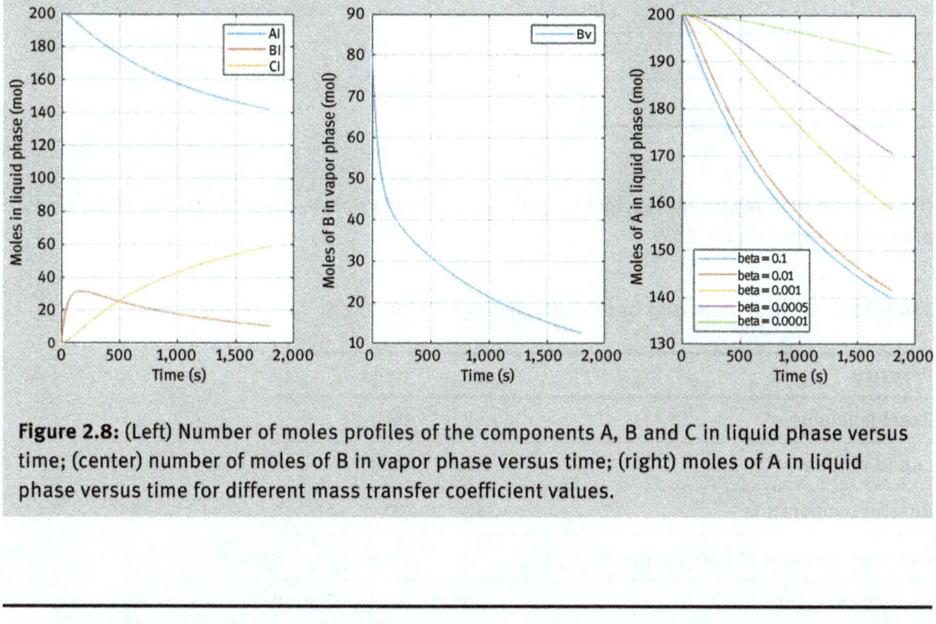

Figure 2.8: (Left) Number of moles profiles of the components A, B and C in liquid phase versus time; (center) number of moles of B in vapor phase versus time; (right) moles of A in liquid phase versus time for different mass transfer coefficient values.

Example 2.6 Gas–liquid single reaction, constant volume, constant pressure, isothermal reactor

A gas–liquid isothermal reactor is operated with a constant pressure in vapor phase with a single chemical reaction described by the following stoichiometry:

$$A + B \rightarrow C$$

The pressure in the gas space in the reactor is maintained constant by means of an automatic control system that supply gaseous reactant B as the reaction consumes it, after the transfer to the liquid phase. Reactant B is initially charged in the reactor and, at time = 0, is present only in the gas phase.

This reactant is gradually transferred to the liquid phase in which the reaction occurs. The kinetic of the reaction is represented by a second-order expression with a constant as in Table 2.10:

Table 2.10: Rate equation and kinetic parameter for Example 2.6.

Reaction	Rate expression	k
1	$r = kC_A C_B$	$1.2e{-}5$ (m^3/(mol s))

The initial charge in the reactor (liquid phase) is summarized in Table 2.11:

Table 2.11: Initial charge in the batch reactor, Example 2.6.

Component	Initial moles (mol)
A	200
B	0
C	0

Other useful data regarding reactor characteristics and physicochemical properties of the mixture are summarized in Table 2.12.

Table 2.12: Physical properties and reactor settings, Example 2.6.

Property	Value	Units
Reactor volume	2	(m^3)
Liquid volume	1.2	(m^3)
Reactor temperature	360	(K)
Constant pressure of B in gas phase	3	(atm)
Henry's constant	20	(mol/(m^3 atm))
Time span for integration	0–3,600	(s)
Volumetric mass transfer coefficient	0.01	(s^{-1})

Part 1

In a first subplot, describe the time profiles of the moles of three components in liquid phase in the time range reported in the previous table. For solving the problem in part 1 and assuming as negligible the gas side mass transfer resistance, the following equations must be solved simultaneously:

$$\frac{dn_A^L}{dt} = V_L(-r) \quad \frac{dn_B^L}{dt} = V_L(-r) + V_L J_B \quad \frac{dn_C^L}{dt} = V_R(+r)$$

$$J_B = \beta(C_B^* - C_B) \quad C_B^* = K_H P_B$$

Part 2

Perform a parametric study by solving the same problem defined in part 1 but with various values assigned to β as in the subsequent list: 0.1; 0.01; 0.001; 0.0005; 0.0001. Report, in a second subplot, the trend of moles of component A in liquid phase for the various values of mass transfer coefficient.

Matlab code for the solution of this example and results (Figure 2.9) are reported as follows:

```
% example 2.6
clc,clear

global k Vr Vl beta T Kh Rgas pB0

%% part 1
Vr=2;                 % overall reactor volume (m3)
Vl=1.2;               % liquid phase volume (m3)
k=1.2e-5;             % kinetic constant (m3/(mol s))
beta=0.01;            % G-L mass transfer coefficient (s)
T=360;                % temperature (K)
Kh=20;                % Henry's constant (mol/(m3 atm))
pB0=3;                % initial pressure of B (atm)
Rgas=0.08205/1000; % gas constant (atm m3/(mol K))
Vv=Vr-Vl;             % gas phase volume (m3)

nAl0=200;             % initial moles of A in liquid (mol)
nBl0=0;               % initial moles of B in liquid (mol)
nCl0=0;               % initial moles of C in liquid (mol)
nBv0=pB0*Vv/(Rgas*T); % moles of B in gas (mol)

n0=[nAl0 nBl0 nCl0];
tspan=0:0.1:3600;

[tx,nx]=ode15s(@ode_ex2_06,tspan,n0);

%% part 2 - parametric study of beta
betax=[0.1 0.01 0.001 0.0005 0.0001];
for j=1:length(betax)
  beta=betax(j);
  [tx,nxx]=ode15s(@ode_ex2_06,tspan,n0);
  na(:,j)=nxx(:,1);
```

```
end

%% plot
subplot(1,2,1)
plot(tx,nx(:,1),tx,nx(:,2),tx,nx(:,3))

grid
xlabel('Time (s)')
ylabel('Moles in liquid phase (mol)')
legend('Al','Bl','Cl')

subplot(1,2,2)
plot(tx,na)
grid
xlabel('Time (s)')
ylabel('Moles of A in liquid phase (mol)')
legend('beta=0.1','beta=0.01','beta=0.001','beta=0.0005','beta=0.0001')

function [dn] = ode_ex2_06(t,n)
global k Vr Vl beta T Kh Rgas pB0

nAl=n(1);
nBl=n(2);
nCl=n(3);

Vv=Vr-Vl;
cA=nAl/Vl;
cB=nBl/Vl;
cC=nCl/Vl;
pB=pB0;
r=k*cA*cB;

cBs=Kh*pB;
J=beta*(cBs-cB);
dn(1)=-Vl*r;
dn(2)=-Vl*r+Vl*J;
dn(3)=+Vl*r;

dn=dn';
end
```

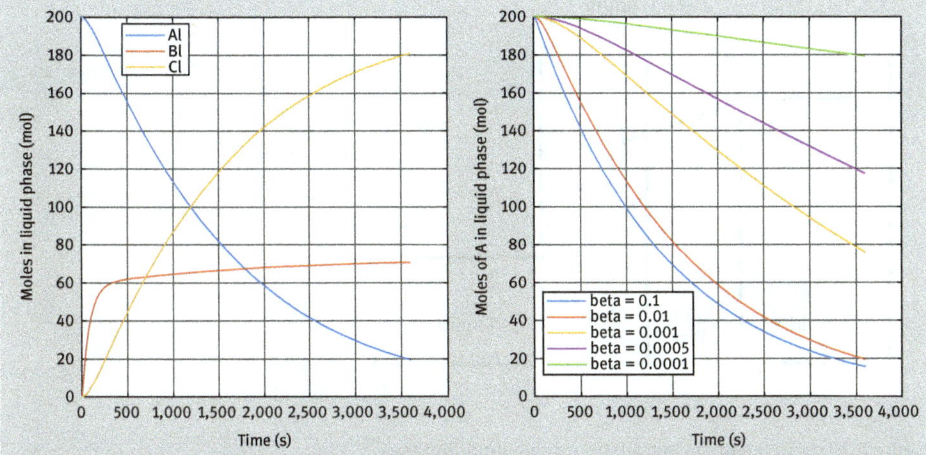

Figure 2.9: (Left) Number of moles profiles of the components A, B and C in liquid phase versus time; (right) moles of A in liquid phase versus time for different mass transfer coefficient values.

2.3.2 Liquid–liquid two-phase reactor

When two immiscible liquids are put in contact in a BR, the mass transfer occurs between two liquid phases. This is particularly relevant if a chemical reaction occurs between some component initially present in liquid 1 (upper phase) and other present in liquid 2 (lower phase). The reaction rate could be, in this case, strongly affected by the liquid–liquid mass transfer phenomenon that is overlapped to chemical reaction [12]. The scheme of mass transfer, adopting also in this case the two-film Withman theory, is reported in Figure 2.10.

The expressions for mass transfer flows, for a generic component i, from bulk of liquid 1 to interface and then from interface to bulk of liquid 2, can be written as

$$\begin{aligned} J_1^i &= \beta_1^i (C_1^i - C_1^{i*}) \\ J_2^i &= \beta_2^i (C_2^{i*} - C_2^i) \end{aligned} \tag{2.25}$$

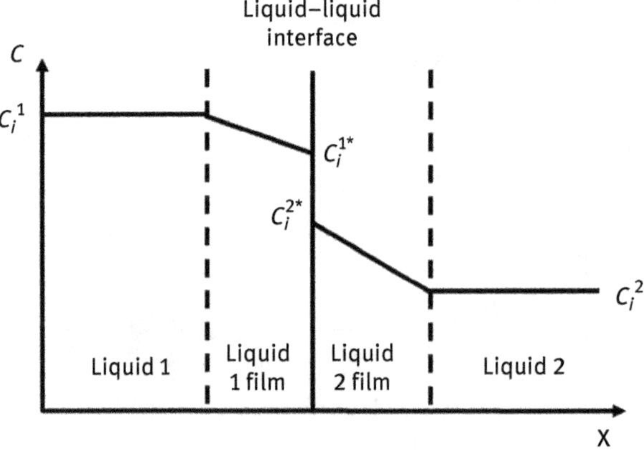

Figure 2.10: Scheme of mass transfer for two liquid phases.

If we introduce the assumption that in correspondence of the liquid–liquid interface, the concentrations of i at both sides are in thermodynamic equilibrium, and they are related to each other by the equilibrium partitioning constant, $H_L{}^i$:

$$H_L^i = \frac{C_1^{i*}}{C_2^{i*}} \tag{2.26}$$

or

$$C_1^{i*} = H_L^i C_2^{i*} \tag{2.27}$$

As the concentration at the separation interface is not easily accessible, we can assume a stationary state in correspondence of this interface (zero accumulation) and solve for the unknown concentration. The steady state at the interface is expressed by the relation:

$$J_1^i V_1 = J_2^i V_2 \tag{2.28}$$

By substituting relations (2.25) and (2.27) into eq. (2.28) and rearranging algebraically, we obtain

$$C_2^{i*} = \frac{V_1 \beta_1^i C_1^i + V_2 \beta_2^i C_2^i}{V_1 \beta_1^i H_L^i + V_2 \beta_2^i} \tag{2.29}$$

In eq. (2.29), the interface concentration of i in liquid 2 is function only of known quantities. After the definition of mass transfer characteristics of the system, mass balance can be developed by referring to the reactor scheme reported in Figure 2.11.

For a detailed derivation of mass balance equations, we assume, generally, that chemical reactions occur in both liquid phases:

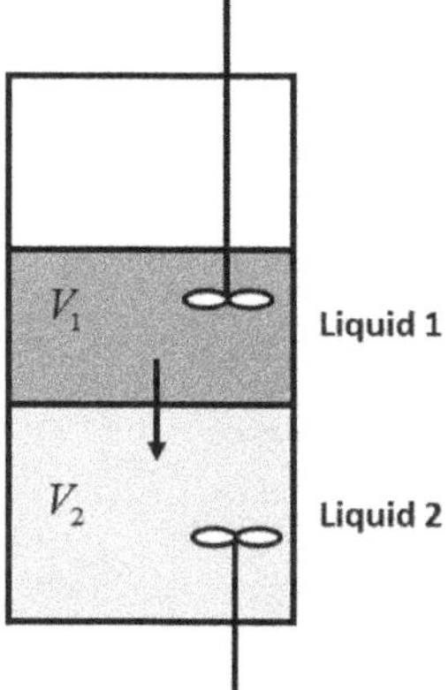

Figure 2.11: Reactor scheme for mass balances.

According to the scheme reported in Figure 2.11, the mass balance for liquid–liquid BR can be developed as follows:
Liquid phase 1

$$\frac{dn_i^1}{dt} = V_1 v_i r_1 - J_i^1 V_1 \tag{2.30}$$

Liquid phase 2

$$\frac{dn_i^2}{dt} = V_2 v_i r_2 + J_i^1 V_1 \tag{2.31}$$

In some cases, the presence of a second immiscible phase is justified for equilibrium reaction that could be difficult to shift toward the products. In this case, the choice of the second phase is crucial and is of particular importance to the affinity (measured by constant H_L) of the immiscible solvent with the component (or components) to be extracted. In this way, by subtracting a product from the reaction environment, the reaction is shifted to the right and the product is recovered in the second phase.

Example 2.7 Liquid–liquid system, single reaction in each phase, constant volume, constant pressure, isothermal reactor

A liquid–liquid isothermal reactor is used to perform the two following chemical reactions occurring in two separated immiscible liquid phases:

$$A + B \rightarrow C + D \text{ (liquid phase 1)}$$
$$B + E \rightarrow C \text{ (liquid phase 2)}$$

The only two components that are partitioned between the two liquid phases are B and C and mass transfer can be described by the two-film theory. The kinetic of the reactions is represented by a second-order expression with a constant as in Table 2.13:

Table 2.13: Rate equation and kinetic parameter for Example 2.7.

Reaction	Rate expression	K_i
1	$r_1 = k_1 C_A^1 C_B^1$	1.2e−5 (m^3/(mol s))
2	$r_2 = k_2 C_B^2 C_E^2$	1.2e−6 (m^3/(mol s))

The initial charge in the reactor (liquid phases 1 and 2) is summarized in Table 2.14:

Table 2.14: Initial charge in the batch reactor, Example 2.7.

Component	Initial moles (mol)
A (liquid 1)	200
B (liquid 1)	500
C (liquid 1)	0
D (liquid 1)	0
B (liquid 2)	0
C (liquid 2)	30
E (liquid 2)	300

Other useful data regarding reactor characteristics and physicochemical properties of the mixture are summarized in Table 2.15.

Table 2.15: Physical properties and reactor settings, Example 2.7.

Property	Value	Units
Volume of liquid phase 1	3	(m^3)
Volume of liquid phase 1	5	(m^3)
Partition constant for B, H_B	0.50	(–)
Partition constant for C, H_C	0.35	(–)
Time span for integration	0–3,600	(s)

Table 2.15 (continued)

Property	Value	Units
Mass transfer coeff. for B side 1, β_1^B	0.01	(s^{-1})
Mass transfer coeff. for B side 2, β_2^B	0.03	(s^{-1})
Mass transfer coeff. for C side 1, β_1^C	0.01	(s^{-1})
Mass transfer coeff. for C side 2, β_2^C	0.03	(s^{-1})

The objective of this example is to build two separate subplots in which the mole profiles are reported for the components in the two phases in equilibrium. The mass balance for this system is represented, first of all, by mass transfer equations for components B and C that are partitioned:

$$J_1^i = \beta_1^i (C_1^i - C_1^{i*}) \quad J_2^i = \beta_2^i (C_2^{i*} - C_2^i) \quad i = B, C$$

$$H_L^i = \frac{C_1^{i*}}{C_2^{i*}} \qquad C_1^{i*} = H_L^i C_2^{i*} \qquad C_2^{i*} = \frac{V_1 \beta_1^i C_1^i + V_2 \beta_2^i C_2^i}{V_1 \beta_1^i H_L^i + V_2 \beta_2^i}$$

These mass transfer equations must be coupled with mass balance differential equations written separately for the two liquid phases in equilibrium:

$$\begin{cases} \dfrac{dn_A^1}{dt} = -V_1 r_1 \\ \dfrac{dn_B^1}{dt} = -V_1 r_1 - J_B^1 V_1 \\ \dfrac{dn_C^1}{dt} = +V_1 r_1 - J_C^1 V_1 \\ \dfrac{dn_D^1}{dt} = +V_1 r_1 \end{cases} \quad \text{Liquid phase 1}$$

$$\begin{cases} \dfrac{dn_B^2}{dt} = -V_2 r_2 + J_B^1 V_1 \\ \dfrac{dn_E^2}{dt} = -V_2 r_2 \\ \dfrac{dn_C^2}{dt} = +V_2 r_2 + J_C^1 V_1 \end{cases} \quad \text{Liquid phase 2}$$

It is worth noting that material balances are solved for mole numbers of each component in the two phases while mass transfer equations contain concentrations. These can be evaluated by dividing mole number by the volume of the related liquid phase.

As a check of the mutual influence between mass transfer and reaction in a liquid–liquid system, build a third plot in which an overall balance on component B is reported as a function of time. From stoichiometric considerations, the balance on B is represented by the following relation:

$$n_B^{01} = n_B^1 + n_B^2 + n_D^1 + (n_E^{02} - n_E^2)$$

Matlab code for the solution of this example and results are reported as follows:

```
% example 2.7
clc,clear

global V1 V2 k1 k2 HLB HLC
global beta1B beta2B beta1C beta2C

V1=3;           % volume of liquid phase 1 (m3)
V2=5;           % volume of liquid phase 2 (m3)
k1=1.2e-5;      % kinetic constant (m3/(mol s))
k2=1.2e-6;      % kinetic constant (m3/(mol s))
HLB=0.50;       % liq-liq partition constant for B (-)
HLC=0.35;       % liq-liq partition constant for C (-)
beta1B=0.01;    % B liq-liq mass transfer coeff. (s^-1)
beta2B=0.03;    % B liq-liq mass transfer coeff. (s^-1)
beta1C=0.01;    % C liq-liq mass transfer coeff. (s^-1)
beta2C=0.03;    % C liq-liq mass transfer coeff. (s^-1)

nA10=200;       % initial moles of A in liquid 1 (mol)
nB10=500;       % initial moles of B in liquid 1 (mol)
nC10=0;         % initial moles of C in liquid 1 (mol)
nD10=0;         % initial moles of D in liquid 1 (mol)

nB20=0;         % initial moles of B in liquid 1 (mol)
nC20=30;        % initial moles of C in liquid 1 (mol)
nE20=300;       % initial moles of e in liquid 1 (mol)

n0=[nA10 nB10 nC10 nD10 nB20 nC20 nE20];
tspan=0:1:3600;

[tx,nx]=ode15s(@ode_ex2_07,tspan,n0);
bal=nB10-nx(:,2)-nx(:,5)-nx(:,4)-(nE20-nx(:,7));

%% plot
subplot(1,3,1)
plot(tx,nx(:,1:4))
grid
xlabel('Time (s)')
ylabel('Moles in liquid phase 1 (mol)')
legend('A','B','C','D')
title('Liquid phase 1')

subplot(1,3,2)
plot(tx,nx(:,5:7))
grid
```

```
xlabel('Time (s)')
ylabel('Moles in liquid phase 2 (mol)')
legend('B','C','E')
title('Liquid phase 2')

subplot(1,3,3)
plot(tx,bal)
grid
xlabel('Time (s)')
ylabel('Balance on B (mol)')
title('Balance')

function [dn] = ode_ex2_07(t,n)
global V1 V2 k1 k2 HLB HLC
global beta1B beta2B beta1C beta2C

nA1=n(1); nB1=n(2); nC1=n(3); nD1=n(4);
nB2=n(5); nC2=n(6); nE2=n(7);

cA1=nA1/V1; cB1=nB1/V1; cC1=nC1/V1; cD1=nD1/V1;
cB2=nB2/V2; cC2=nC2/V2; cE2=nE2/V2;

r1=k1*cA1*cB1;
r2=k2*cB2*cE2;

cB2s=(V1*beta1B*cB1+V2*beta2B*cB2)/(V1*beta1B*HLB+V2*beta2B);
cB1s=cB2s*HLB;
cC2s=(V1*beta1C*cC1+V2*beta2C*cC2)/(V1*beta1C*HLC+V2*beta2C);
cC1s=cC2s*HLC;

J1B = beta1B*(cB1-cB1s);
J2B = beta2B*(cB2s-cB2);
J1C = beta1C*(cC1-cC1s);
J2C = beta2C*(cC2s-cC2);

% liquid phase 1
dn(1) = -V1*r1;
dn(2) = -V1*r1 - V1*J1B;
dn(3) = +V1*r1 - V1*J1C;
dn(4) = +V1*r1;

% liquid phase 2
dn(5) = -V2*r2 + V2*J2B;
dn(6) = +V2*r2 + V2*J2C;
dn(7) = -V2*r2;

dn=dn';
end
```

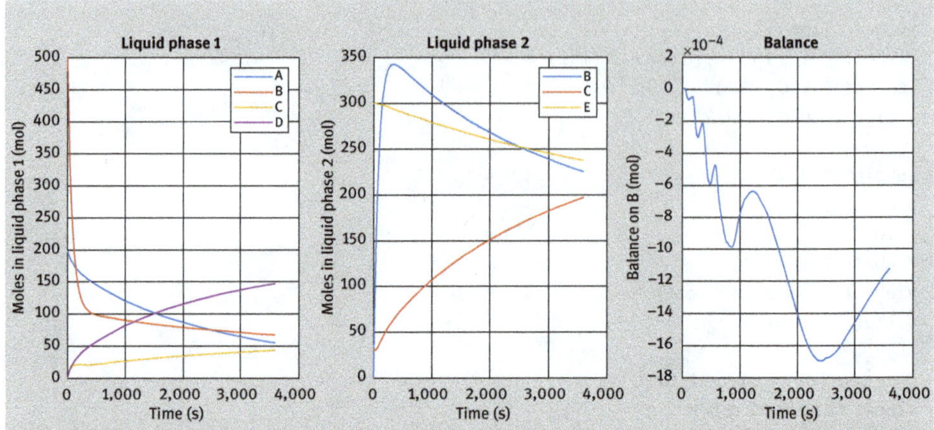

Figure 2.12: (Left) Number of mole profiles of components A, B, C and D in liquid phase 1 versus time; (center) number of moles of B, C and E in liquid phase 2 versus time; (right) overall mass balance of B component versus time.

Example 2.8 Liquid–liquid system, single equilibrium reaction in one phase, second phase used as extracting medium, constant volume, constant pressure, isothermal reactor

An equilibrium reaction occurs in a liquid phase 1 with the following stoichiometry:

$$A + B \longleftrightarrow C + D \text{ (liquid phase 1)}$$

The extent of reaction is limited by chemical equilibrium that prevents a complete conversion of reactants into products. An interesting possibility to shift the reaction to the right, and consequently obtain a full conversion, is to introduce in the system a second liquid phase (solvent) immiscible with the previous one and with high affinity with respect to one of the products. In this way, one or more products are removed from the reaction environment and the reaction is shifted to the right. In our example, we assume that the only component that is partitioned between the two phases is C and its mass transfer can be described by the two-film theory. The kinetic of the reactions is represented by a second-order equilibrium expression with a constant as in Table 2.16:

Table 2.16: Rate equation and kinetic parameter for Example 2.8.

Reaction	Rate expression	Constants
1	$r_1 = k_1 C_A^1 C_B^1 \left(1 - \frac{1}{K_e} \frac{C_C^1 C_D^1}{C_A^1 C_B^1}\right)$	$k_1 = 3.2e{-5}$ (m^3/(mol s)) $K_e = 1.2$ (–)

The initial charge in the reactor (liquid phases 1 and 2) is summarized in Table 2.17:

Table 2.17: Initial charge in the batch reactor, Example 2.8.

Component	Initial moles (mol)
A (liquid 1)	200
B (liquid 1)	350
C (liquid 1)	0
D (liquid 1)	20
C (liquid 2)	0

Other useful data regarding reactor characteristics and physicochemical properties of the mixture are summarized in Table 2.18.

Table 2.18: Physical properties and reactor settings, Example 2.8.

Property	Value	Units
Volume of liquid phase 1	5	(m^3)
Volume of liquid phase 1	20	(m^3)
Part 1 – partition constant for C, H_C	200	(–)
Part 2 – partition constant for C, H_C	0.02	(–)
Time span for integration	0–4,000	(s)
Mass transfer coeff. for C side 1, β_1^C	0.01	(s^{-1})
Mass transfer coeff. for C side 2, β_2^C	0.03	(s^{-1})

Part 1

In a first simulation, the objective is the description of the system in reaching the equilibrium conditions. This can be done by adopting for the partition constant, related to component C, a relatively high value that involves low concentration in the extracting solvent. Alternatively, this situation can be represented by assuming a negligible or infinitesimal volume for the phase 2. In our example, the first approach is used, and $H_C = 200$.

The mass transfer for component C, that is, the unique partitioned substance, is described by the following relations:

$$J_1^C = \beta_1^C (C_1^C - C_1^{C^*}) \qquad J_2^C = \beta_2^C (C_2^{C^*} - C_2^C)$$

$$H_L^C = \frac{C_1^{C^*}}{C_2^{C^*}} \qquad C_1^{C^*} = H_L^C C_2^{C^*} \qquad C_2^{C^*} = \frac{V_1 \beta_1^C C_1^C + V_2 \beta_2^C C_2^C}{V_1 \beta_1^C H_L^C + V_2 \beta_2^C}$$

These mass transfer equations must be coupled with mass balance differential equations written separately for the two liquid phases in equilibrium:

$$\begin{cases} \dfrac{dn_A^1}{dt} = -V_1 r_1 \\ \dfrac{dn_B^1}{dt} = -V_1 r_1 \\ \dfrac{dn_C^1}{dt} = +V_1 r_1 - J_C^1 V_1 \\ \dfrac{dn_D^1}{dt} = +V_1 r_1 \end{cases} \qquad \text{Liquid phase 1}$$

$$\frac{dn_C^2}{dt} = +J_C^1 V_1 \qquad \text{Liquid phase 2}$$

In this first part, build three plots reporting, respectively, the moles in liquid phase1, moles of C in phase 2 and the conversion of component A.

Part 2

Repeat the calculations described in part 1 but using a partition constant $H_C = 0.02$, in order to favor the extraction of component C in the solvent phase.

Matlab code for the solution of this example and results are reported as follows:

```
% example 2.8
clc,clear
global k Ke beta1 beta2 HC
global V1 V2
V1=5;           % volume of liquid phase 1
V2=30;          % volume of liquid phase 2
k=3.2e-5;       % kinetic constant
Ke=1.2;         % equilibrium constant
beta1=0.1;      % mass transfer coefficient side 1
beta2=0.3;      % mass transfer coefficient side 2
nA10=200;       % initial moles of A in liquid 1
nB10=350;       % initial moles of B in liquid 1
nC10=0;         % initial moles of C in liquid 1
nD10=20;        % initial moles of D in liquid 1
nC20=0;         % initial moles of C in liquid 2
nAtot0=nA10;
n0=[nA10 nB10 nC10 nD10 nC20];
tspan=0:4000;
%% part 1
HC=200;         % liq-liq partition constant for C
[tx,nx1]=ode15s(@ode_ex2_08,tspan,n0);
nAx=nx1(:,1);
xA1=(nAtot0-nAx)/nAtot0;

%% part 2
HC=0.02;        % liq-liq partition constant for C
[tx,nx2]=ode15s(@ode_ex2_08,tspan,n0);
nAx=nx2(:,1);
```

```
xA2=(nAtot0-nAx)/nAtot0;
%% plots
subplot(2,3,1)
plot(tx,nx1(:,1:4))
grid
xlabel('Time (s)')
ylabel('Moles (mol)')
title('(a) Liquid phase 1')
legend('A','B','C','D')
subplot(2,3,2)
plot(tx,nx1(:,5))
grid
xlabel('Time (s)')
ylabel('Moles (mol)')
title('(b) Liquid phase 2')
legend('C')

subplot(2,3,3)
plot(tx,xA1)
grid
xlabel('Time (s)')
ylabel('Conversion of A (-)')
title('(c) Conversion of A')

subplot(2,3,4)
plot(tx,nx2(:,1:4))
grid
xlabel('Time (s)')
ylabel('Moles (mol)')
title('(d) Liquid phase 1')
legend('A','B','C','D')

subplot(2,3,5)
plot(tx,nx2(:,5))
grid
xlabel('Time (s)')
ylabel('Moles (mol)')
title('(e) Liquid phase 2')
legend('C')

subplot(2,3,6)
plot(tx,xA2)
grid
xlabel('Time (s)')
ylabel('Conversion of A (-)')
title('(f) Conversion of A')
function [dn] = ode_ex2_08(t,n)
global k Ke beta1 beta2 HC
global V1 V2
```

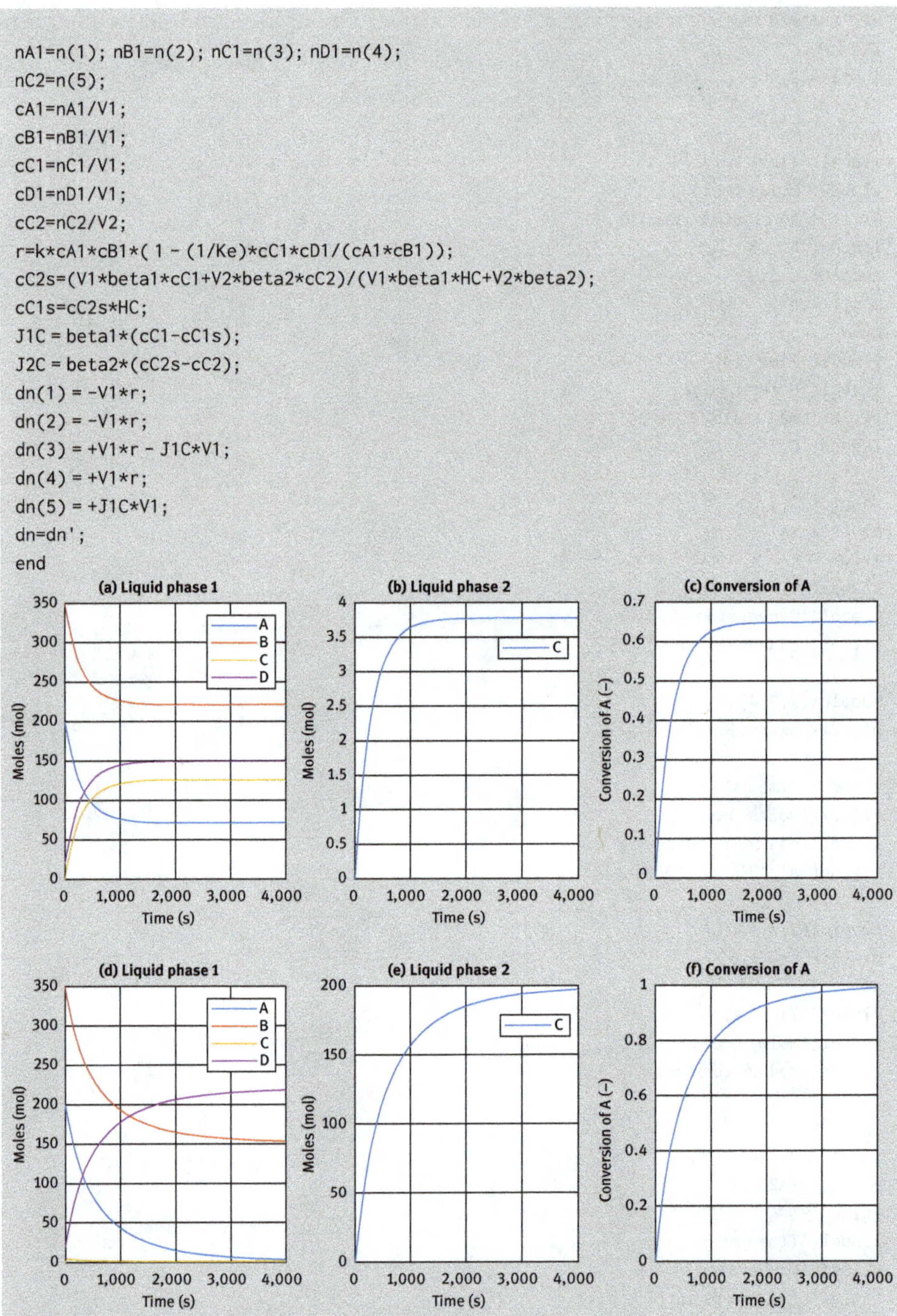

```
nA1=n(1); nB1=n(2); nC1=n(3); nD1=n(4);
nC2=n(5);
cA1=nA1/V1;
cB1=nB1/V1;
cC1=nC1/V1;
cD1=nD1/V1;
cC2=nC2/V2;
r=k*cA1*cB1*( 1 - (1/Ke)*cC1*cD1/(cA1*cB1));
cC2s=(V1*beta1*cC1+V2*beta2*cC2)/(V1*beta1*HC+V2*beta2);
cC1s=cC2s*HC;
J1C = beta1*(cC1-cC1s);
J2C = beta2*(cC2s-cC2);
dn(1) = -V1*r;
dn(2) = -V1*r;
dn(3) = +V1*r - J1C*V1;
dn(4) = +V1*r;
dn(5) = +J1C*V1;
dn=dn';
end
```


Figure 2.13: (a,d) Number of moles of components A, B, C and D in liquid phase 1 versus time; (b,e) number of moles of C in liquid phase 2 versus time and (c,f) fractional conversion of A.

In Figure 2.13, (a), (b) and (c) are related to part 1 of the example while the remaining (d), (e) and (f) are related to part 2.

2.4 Fed-batch reactors

A variation in the basic configuration of BR is represented by fed-batch reactor, and its scheme is reported in Figure 2.14. This setup is characterized by a liquid or gas feed that is added along the time to the initially charged mixture.

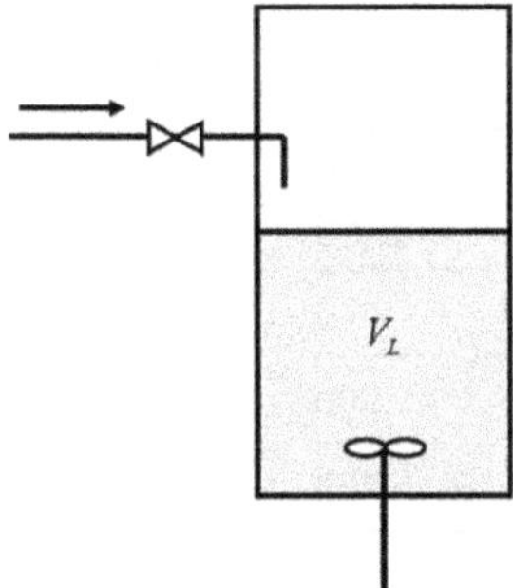

Figure 2.14: Scheme of a fed-batch reactor.

The reaction volume increases until a maximum holdup level is reached, and the feed is then stopped. The amount of feed stream could be a function of time (ramp, step, rectangular pulse, etc.), but in all the cases, a control system must be operative. For example, if the reactions occur in liquid phase, a level control (high-level alarm) must be provided. On the contrary, for a gas-phase reactive system, the reactor must be equipped with a pressure control (high-pressure alarm). These two types of controls act on the feed stream in order to avoid, respectively, overload and overpressure of the apparatus. In the case of a liquid-phase reaction and a liquid feed that is a function of time, the mass balance for each component in the system is represented by the following relation:

$$\frac{dn_i}{dt} = V_R \nu_i r + F_i \tag{2.32}$$

The application of material balance equations (2.32) is strictly related to control strategy according to which the feed F_i is modulated or stopped when the liquid volume in the reactor reaches a maximum allowable level.

Example 2.9 Liquid-phase single reaction, variable volume, constant pressure, isothermal fed-batch reactor

In a fed-batch reactor, a single reaction occurs with the following stoichiometry:

$$A + B \rightarrow C$$

This reaction is characterized by a second-order kinetics with a reaction rate expression expressed by the relation ($k = 1e{-}6\ m^3/(mol\ s)$):

$$r = kC_A C_B$$

The properties of the three components and the initial charge in the reactor are summarized in Table 2.19:

Table 2.19: Initial charge in the reactor, molar flow rates and properties for the fed-batch reactor of Example 2.9.

Component	Initial moles (mol)	Molar flow rate in feed (mol/s)	Molecular weight (kg/mol)	Density (kg/m³)
A	1,000	8	0.028	570
B	1,200	6	0.018	1,000
C	0	0	0.046	789

The reactor is operated with an inlet feed and a control in maximum allowable volume of the liquid phase that is set at $V_{max} = 0.2\ m^3$. The feed is constituted by the molar flow rates of the components as reported in the previous table.

To describe the time profiles of both components' concentration and liquid volume, the following equation must be solved:

$$\frac{dn_A}{dt} = -V_R r + F_A \quad \frac{dn_B}{dt} = -V_R r + F_B \quad \frac{dn_C}{dt} = +V_R r + F_C$$

$$V_R = \sum_{i=1}^{N_C} \frac{n_i M_{Wi}}{\rho_i} \quad C_i = \frac{n_i}{V_R}$$

Matlab code for the solution of this example and results (Figure 2.15) are reported as follows:

```
% example 2.9
clc,clear
global mw ro ni k F Vmax
mw=[0.028 0.018 0.046]; % molecular weights (kg/mol)
ro=[ 570  1000  789];    % densities (kg/m3)
ni=[ -1  -1     1];      % stoich. coefficients

n0=[1000 1200 0];        % initial moles (mol)
tspan=0:0.1:800;         % time span (s)
```

```
k=1e-6;                % kinetic constant (m3/(mol s))
F=[8 6 0];             % feed (mol/s)
Vmax=0.2;              % maximum volume (m3)

options=odeset('MaxStep',0.1);
[tx,nx]=ode45(@ode_ex2_09,tspan,n0,options);
for j=1:length(nx)
  Vr(j)=0;
  for k=1:3
    Vr(j)=Vr(j)+nx(j,k)*mw(k)/ro(k);
  end
end

%% plot
subplot(1,2,1)
plot(tx,nx(:,1),tx,nx(:,2),tx,nx(:,3))
grid
xlabel('Time (s)')
ylabel('Moles (mol)')
legend('A','B','C')

subplot(1,2,2)
plot(tx,Vr)
grid
xlabel('Time (s)')
ylabel('Reaction volume (m3)')

function [dn] = ode_ex2_09(t,n)
global mw ro ni k F Vmax

Vr=sum(n'.*mw./ro);     % reaction volume (m3)
c=n/Vr;                 % concentrations (mol/m3)
r=k*c(1)*c(2);          % reaction rate (mol/(m3 s))

if Vr<Vmax     % check if Vmax is reached
  Fx=F;
else
  Fx=[0 0 0];
end
dn(1)=ni(1)*Vr*r + Fx(1);
dn(2)=ni(2)*Vr*r + Fx(2);
dn(3)=ni(3)*Vr*r + Fx(3);
dn=dn';
end
```

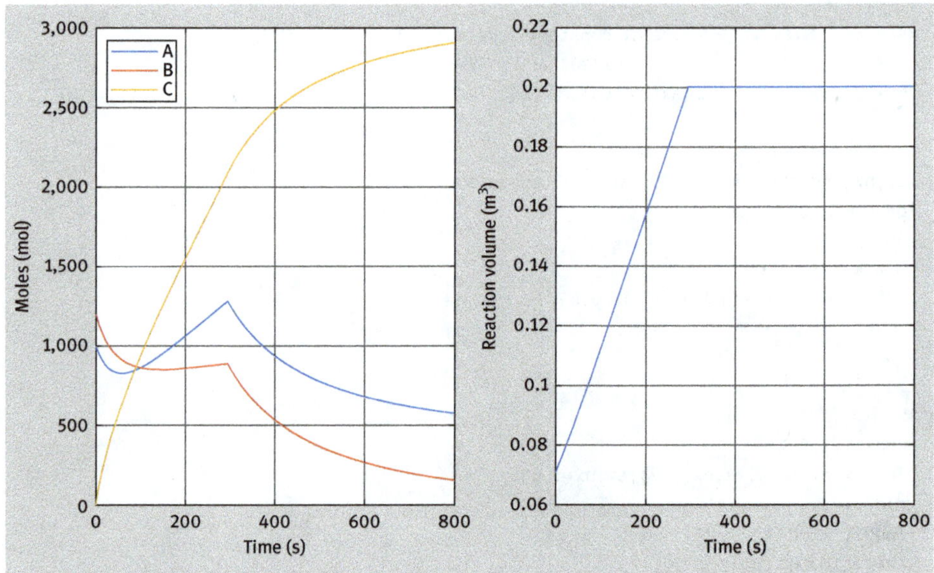

Figure 2.15: (Left) Number of moles of the components A, B and C in liquid phase versus time; (right) reaction volume versus time.

List of symbols

a	Liquid–liquid specific interface area	$[m^2/m^3]$
n_i	Moles of component *i*	[mol]
n_i^G	Moles of component *i* in gas phase	[mol]
t	Time	[s]
C_i, C_j	Concentration of component *i* or *j*	$[mol/m^3]$
C_i^1, C_i^2	Concentration of component *i* in liquid phases 1 and 2	$[mol/m^3]$
C_i^{1*}	Equilibrium concentration of component *i* in liquid phases 1	$[mol/m^3]$
C_i^{2*}	Equilibrium concentration of component *i* in liquid phases 2	$[mol/m^3]$
C_i^*	Equilibrium concentration of component *i*	$[mol/m^3]$
P_i^*	Equilibrium pressure of component *i*	[atm]
C_{pm}	Average specific heat of reaction mixture	[J/(kg K)]
ΔH_r	Heat of reaction	[J/mol]
ΔH_{rj}	Heat of reaction *j*	[J/mol]
Q_E	Heat exchanged from the reactor to surroundings	[J/s]
U	Overall heat transfer coefficient	$[J/(m^2 s K)]$
A	Heat transfer area	$[m^2]$
E_A, E_{Ai}	Activation energy, activation of *i*th reaction	[J/mol]
k_i	Kinetic constant of *i*th chemical reaction	[*]
k_{iref}	Reference kinetic constant of *i*th chemical reaction	[*]
k_{Li}	Gas–liquid mass transfer coefficient of component *i*	[m/s]

m	Mass	[kg]
H_i	Gas–liquid partition constant (Henry's constant)	[atm m^3/mol]
H_L^i	Liquid–liquid partition constant	[–]
R	Gas constant	[atm m^3/(mol K)]
V_R	Reactor volume	[m^3]
V_L	Liquid phase volume	[m^3]
V_G	Gas phase volume	[m^3]
V_1	Volume of immiscible liquid phase 1	[m^3]
V_2	Volume of immiscible liquid phase 2	[m^3]
V_j	Volume of component j	[m^3]
r	Reaction rate	[mol/(m^3 s)]
P	Total pressure	[atm]
p_j, P_i	Partial pressure of component i or j	[atm]
y_i	Mole fraction of component i	[–]
r_k	Reaction rate of reaction k	[mol/(m^3 s)]
R_i	Overall generation/consumption rate of component i	[mol/(m^3 s)]
Mw_j	Molecular weight of component j	[kg/mol]
N_C	Number of components in the mixture	[–]
N_{Ci}	Number of components involved in the reaction i	[–]
N_R	Number of independent chemical reactions	[–]
T	Temperature of the reaction mixture	[K]
T_S	Temperature of heating/cooling fluid	[K]
T_{ref}	Reference temperature in Arrhenius equation	[K]
J_i	Gas–liquid mass transfer flow of component i	[mol/(m^3 s)]
J_1^i	Liquid–liquid mass transfer flow of component i (bulk 2 to interface)	[mol/(m^3 s)]
J_2^i	Liquid–liquid mass transfer flow of component i (interface to bulk 1)	[mol/(m^3 s)]
F_i	Molar flow rate of component i in feed	[mol/s]
ρ_j	Density of component j	[kg/m^3]
ρ_m	Density of mixture	[kg/m^3]
λ_j	Reaction order for component j	[–]
β_i	Volumetric gas–liquid mass transfer coefficient for component i	[s^{-1}]
β_1^i	Volumetric gas–liquid mass transfer coefficient for component i in phase 1	[s^{-1}]
β_2^i	Volumetric gas–liquid mass transfer coefficient for component i in phase 2	[s^{-1}]
ν_i	Stoichiometric coefficient of component i	[–]
$\nu_{i,k}$	Stoichiometric coefficient of component i in reaction k	[–]

*Units of kinetic constants depend on the expression of the reaction rate.

References

[1] H.S. Fogler. Elements of Chemical Reaction Engineering (5th edition). Prentice Hall: 2016.

[2] E.B. Nauman. Chemical Reactor Design, Optimization, and Scaleup (2nd edition). Wiley: 2008.

[3] O. Levenspiel Chemical Reaction Engineering (3rd edition). Wiley: 1998.

[4] Ullmann's Encyclopedia of Industrial Chemistry. Wiley: 2007.

[5] V.G. Pangarkar. Design of Multiphase Reactors. John Wiley & Sons: 2014.

[6] G. Astarita. Mass Transfer with Chemical Reaction. Elsevier: 1967.

[7] W.G. Whitman. The two-film theory of gas adsorption. Chemical and Metallurgical Engineering 1932, 29(4), 146–148.
[8] Y.A. Cengel, A.J. Ghajar. Heat and Mass Transfer, Fundamentals and Applications (5th edition). McGraw-Hill: 2015.
[9] T. Salmi, V. Russo. Modelling of a liquid-liquid-solid-gas system: Hydrogenation of dispersed liquid sodium to sodium hydride. Chemical Engineering Journal 2019, 356(15), 445–452.
[10] T. Salmi, V. Russo. Reaction engineering approach to the synthesis of sodium borohydride. Chemical Engineering Science 2019, 199, 79–87.
[11] R. Vitiello, V. Russo, R. Turco, R. Tesser, M. Di Serio, E. Santacesaria. Glycerol chlorination in a gas-liquid semibatch reactor: new catalysts for chlorohydrin production. Chinese Journal of Catalysis 2014, 35(5), 663–669.
[12] E. Santacesaria, R. Turco, M. Tortorelli, V. Russo, M. Di Serio, R. Tesser. Biodiesel process intensification by using static mixers tubular reactors. Industrial & Engineering Chemistry Research 2012, 51, 8777–8787.

Chapter 3
Batch reactors for heterogeneous catalysis

3.1 Introduction

A chemical process is often catalyzed by a heterogeneous catalyst [1, 2]. When a solid component, that is, the catalyst, is put in contact with a fluid phase containing the reactants, a complex reaction pathway describes the physics and the chemistry of the network. In general, six main processes can be individuated considering the fluid bulk phase always well mixed: (i) diffusion in a stagnant fluid film surrounding the particle; (ii) diffusion inside the catalyst pores (intraparticle diffusion); (ii) adsorption of the reactants on the catalyst surface; (iii) chemical reaction; (iv) desorption of the products; (v) back diffusion in the catalyst pores; and (vi) back diffusion in the stagnant fluid film. It is evident that the numerical solution of the main mass and energy balances equations useful to describe the network can be demanding, as concentrations and temperature are dependent on two variables, that is, time and the radius of the particle. In this chapter, the problem will be faced treating two cases of increasing complexity: (i) a single particle model, where the bulk fluid phase will be considered unchanging with time; (ii) fully comprehensive fluid–solid batch reactor model.

3.2 Single particle modeling

The topic of this chapter is the treatment of the special case where a single particle is put in contact with a bulk fluid phase of infinite volume at high concentration. Thus, the occurrence of the chemical reaction would lead only to a slight change in the fluid phase composition and temperature. Different cases will be considered, switching the rate-determining step of the entire process to either the fluid film or the intraparticle region. Finally, both processes will be considered in series.

3.2.1 Fluid–solid external mass transfer

If the intraparticle diffusion limitation can be considered negligible or in the case of a nonporous catalyst, the chemical reaction occurs only in the catalyst surface. In this case, the physics and the chemistry of the process can be described as the occurrence in series of two different phenomena: (i) the diffusion of the reactants in the stagnant fluid film surrounding the catalyst particle; (ii) the chemical reaction of the catalyst surface. The sketch of the presented phenomena is depicted in Figure 3.1.

https://doi.org/10.1515/9783110632927-003

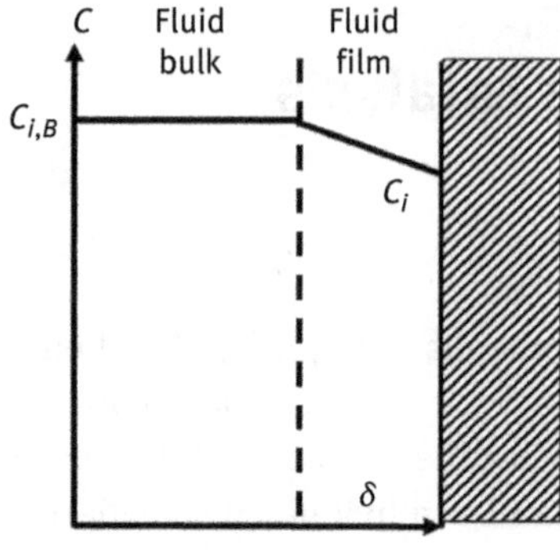

Figure 3.1: Mass transfer in the stagnant film of fluid surrounding the catalyst particle.

To describe the system, the molar balance for the ith component can be derived from the general mass conservation equation that can be written as follows:

$$[\text{accumulation}_i] = [\text{inlet}_i] - [\text{outlet}_i] + [\text{diffusion}_i] + [\text{generation}_i] \tag{3.1}$$

For a batch reactor, no inlet and outlet terms are present. Moreover, for this preliminary case, the accumulation term is neglected as the intent is to focus the attention on the relative importance of the reactive term over the diffusion one. The diffusion term can be written by means of Fick's law.

In an explicit form, eq. (3.1) can be written as follows:

$$J_i = -\nu_i r_s \rightarrow k_m(C_{i,B} - C_i) = kC_i \tag{3.2}$$

$C_{i,B}$ and C_i are the concentrations of the ith component, respectively, in the bulk and on the catalyst surface, k_m is the mass transfer coefficient, k the kinetic constant and r the reaction rate given for catalyst surface units [mol/(m^2 s)].

The concentration in the fluid film phase can be calculated from (3.2):

$$C_i = \frac{k_m}{k_m + k} C_{i,B} \tag{3.3}$$

Thus, the effective rate of the process is given by

$$r_s = kC_i = \frac{kk_m}{k_m + k} C_{i,B} \tag{3.4}$$

The problem can be generalized including the occurrence of several chemical reactions:

$$R_{i,s} = \sum_{k=1}^{Nr} \nu_{i,k} r_{k,s} \tag{3.5}$$

where Nr is the number of independent chemical reactions, $R_{i,s}$ is the overall generation/consumption term for component i, $r_{k,s}$ is the kth reaction rate and $\nu_{i,k}$ is the stoichiometric coefficient of component i in the reaction k (frequently referred to as stoichiometric matrix).

The most uncertain parameter to be defined is surely the mass transfer coefficient, as it is rather hard to be measured for each specific case. Different correlations are proposed in the literature [1], but the common approach is to express it by means of dimensionless numbers:

$$\mathrm{Sh} = \frac{k_m d_p}{D_i} = 2 + 0.6\mathrm{Re}^{1/2}\mathrm{Sc}^{1/3} \tag{3.6}$$

Substituting the Reynolds and Schmidt number definitions, we obtain

$$k_m = 0.6\left(\frac{D_i}{d_p}\right)\mathrm{Re}^{1/2}\mathrm{Sc}^{1/3} = 0.6\left(\frac{D_i}{d_p}\right)\left(\frac{Ud_p}{\nu}\right)^{1/2}\left(\frac{\nu}{D_i}\right)^{1/3} = 0.6\frac{D_i^{2/3}}{\nu^{1/6}}\frac{U^{1/2}}{d_p^{\ 1/2}} \tag{3.7}$$

The mass transfer coefficients depend on two terms related to either physical properties (i.e., viscosity, ν, and molecular diffusivity, D_i) or to operative conditions (particle dimensions, d_p, and fluid velocity, U). General considerations can be drawn:

- Molecular diffusivity always increases with temperature.
- Viscosity is a function of $T^{3/2}$ for gases and T^{-1} for liquids
- It is possible to increase the k_m value by decreasing the particle dimensions or increasing the fluid velocity, that means the stirring rate of the system.

Depending on the case, the rate-determining step of the system can be either the chemical reaction or the mass transfer resistance.

3.2.2 Intraparticle mass and heat transfer

Usually, the industrial catalysts are particles characterized by a relatively high porosity, thus the reactants/products molecules can diffuse in the catalyst pores and there reacts. In this case, the physics of the systems becomes more complex as intraparticle diffusion must be considered (Figure 3.2).

To describe the system, we firstly assume a spherical particle. The molar balance for the ith component can be derived from the general mass conservation equation that can be written as follows:

$$[\mathrm{accumulation}_i] = [\mathrm{inlet}_i] - [\mathrm{outlet}_i] + [\mathrm{diffusion}_i] + [\mathrm{generation}_i] \tag{3.8}$$

For a batch reactor, no inlet and outlet terms are present. Sketching the single particle as in Figure 3.3, each term can be written easily for a spherical shell of a particle, with a surface equal to $4\pi r_p \partial r_p$.

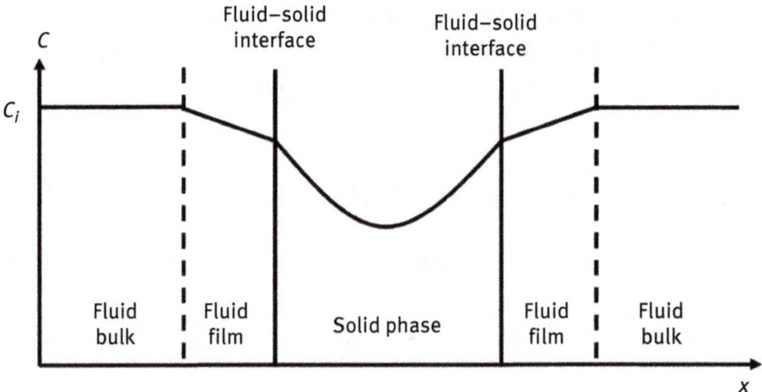

Figure 3.2: Mass transfer in fluid–solid system for a single particle case.

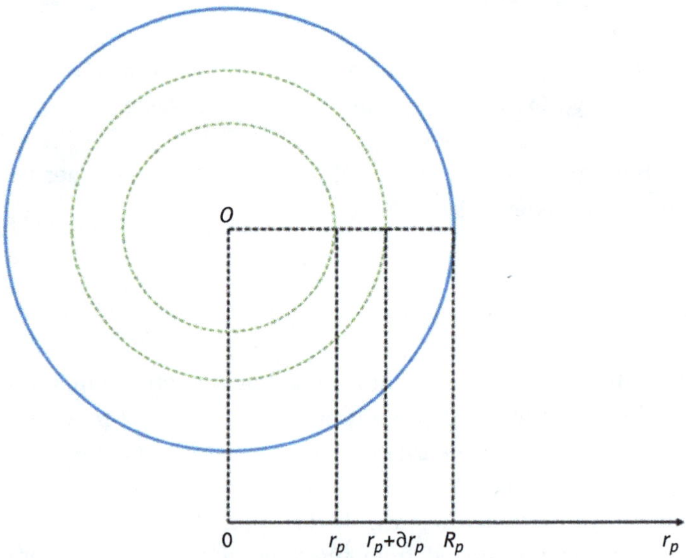

Figure 3.3: Spherical catalyst particle: integration domains.

The explicit form of the mass balance is given as

$$\varepsilon \frac{\partial C_i}{\partial t} = r_p^{-2} \frac{\partial}{\partial r_p}\left(D_{e,i} r_p^2 \frac{\partial C_i}{\partial r_p}\right) + \sum_{k=1}^{N_R} v_{ik} r_k \rho_p \tag{3.9}$$

where ε is the particle porosity, ρ_p the particle density, $D_{e,i}$ the effective diffusivity estimated by weighting the molecular diffusivity by the particle porosity, tortuosity (τ) and the constriction factor (σ):

$$D_{e,i} = \frac{\varepsilon\sigma}{\tau} D_i \tag{3.10}$$

It is evident that the effective diffusivity is always lower than the molecular one, very often 1/10th of it [3].

Equation (3.9) can be rewritten in dimensionless form, defining the dimensionless particle radius coordinate: $x = r_p/R_p$, obtaining:

$$\varepsilon\frac{\partial C_i}{\partial t} = x^{-2}\frac{\partial}{\partial x}\left(\frac{D_{e,i}}{R_p^2}x^2\frac{\partial C_i}{\partial x}\right) + \sum_{k=1}^{N_R} \nu_{ik} r_k \rho_p \tag{3.11}$$

For a particle of general geometry, the mass balance becomes

$$\varepsilon\frac{\partial C_i}{\partial t} = x^{-s}\frac{\partial}{\partial x}\left(\frac{D_{e,i}}{R_p^s}x^s\frac{\partial C_i}{\partial x}\right) + \sum_{k=1}^{N_R} \nu_{ik} r_k \rho_p \tag{3.12}$$

where s is the shape factor, defined as

$$s + 1 = \frac{A}{V} L \tag{3.13}$$

where $s = 0$ for slabs, $s = 1$ for cylindrical and $s = 2$ for spherical particles. Noninteger values and values greater than 2 can be obtained for real particles characterized by high values of the effective surface.

Equation (3.12) represents a partial differential equation (PDE) system, where the concentration of each component diffusing within the catalyst particle bust be solved along time and radial coordinate. Two boundary conditions are needed for, respectively, the surface and the center of the particle:

$$\frac{\partial C_i}{\partial x}\Big|_{x=0} = 0 \tag{3.14}$$

$$\frac{D_{e,i}}{R_p}\frac{\partial C_i}{\partial x}\Big|_{x=1} = k_m\left(C_{i,B} - C_i|_{x=1}\right) \tag{3.15}$$

Equation (3.13) represents the symmetry condition for the parabolic profile of the concentration at the center of the particle. Equation (3.14) expresses the continuity between the fluid bulk and the catalyst surface. If the fluid–solid mass transfer resistance can be neglected, a further simplification rises

$$C_i|_{x=1} = C_{i,Bulk} \tag{3.16}$$

In a similar way, the energy balances can be written, obtaining in dimensionless form, for a general shaped particle:

$$\left(c_{p,p}\rho_p(1-\varepsilon)+\sum_{i=1}^{N_C}c_{p,i}C_i\varepsilon\right)\frac{\partial T}{\partial t}=x^{-s}\frac{\partial}{\partial x}\left(\frac{k_{T,p}}{R_p^s}\frac{\partial T}{\partial x}x^s\right)+\sum_{k=1}^{N_R}(-\Delta_r H_k)\nu_{ik}r_k\rho_p \quad (3.17)$$

In this case, the boundary conditions can be written for, respectively, the center and the surface of the particle as

$$\left.\frac{\partial T}{\partial x}\right|_{x=0}=0 \quad (3.18)$$

$$\left.\frac{k_{T,p}}{R_p}\frac{\partial T}{\partial x}\right|_{x=1}=h\left(T_B-T|_{x=1}\right) \quad (3.19)$$

Or alternatively for the surface, in the case of negligible heat transfer resistance in the fluid film:

$$T|_{x=1}=T_B \quad (3.20)$$

Mass and heat transfer coefficients in the fluid film phases can be related through the Chilton–Colburn analogy, decreasing the number of unknown parameters:

$$\frac{\mathrm{Nu}}{\mathrm{Re}\times\mathrm{Pr}^{1/3}}=\frac{\mathrm{Sh}}{\mathrm{Re}\times\mathrm{Sc}^{1/3}} \quad (3.21)$$

$$\frac{h}{k_{T,f}}\left(\frac{k_{T,f}}{\mu c_{p,f}}\right)^{1/3}=\frac{k_m}{D_e}\left(\frac{\rho D_e}{\mu}\right)^{1/3} \quad (3.22)$$

$$h=\frac{k_m}{D_e}\left(\frac{\rho D_e c_{p,f}}{k_{T,f}}\right)^{1/3}k_{T,f} \quad (3.23)$$

To analyze the influence of the intraparticle mass transfer limitations, two different dimensionless numbers were introduced: the Thiele modulus and the effectiveness factor.

In particular, the Thiele modulus was defined as the ratio between the surface reaction and the diffusion rate of the reactant A, and that for an nth order reaction is defined as follows:

$$\phi_n^2=\frac{kR_p{}^2\rho_p C_{A,B}^{n-1}}{D_{e,A}}=\frac{kR_p\rho_p C_{A,B}^n}{D_{e,A}\left[(C_{A,B}-0)/R_p\right]}=\frac{\text{"A" surface reaction}}{\text{"A" diffusion rate}} \quad (3.24)$$

That, more classically, is reported to be

$$\phi_n=R_p\sqrt{\frac{k\rho_p C_{A,B}^{n-1}}{D_{e,A}}} \quad (3.25)$$

A large value of this quantity corresponds to high intraparticle diffusion limitation, while a small value, the chemical reaction, can be considered the rate-determining step.

The effectiveness factor, instead, is defined as the ratio between the overall reaction rate divided by the reaction rate that would occur only at the surface:

$$\eta = \frac{\int_{x=0}^{x=1} r(4\pi x^2 R_p^2)dx}{\int_{x=0}^{x=1} r|_{x=1}(4\pi x^2 R_p^2)dx} \tag{3.26}$$

It is evident that in the case of negligible intraparticle mass transfer limitations, the effectiveness factor is equal to one, while it is lower than one if concentration gradients are expected. Special cases allow to the derivation of effectiveness factor values greater than one, where the rate inside the catalyst is greater than the reaction rate at the surface. This case is generally present for very exothermic reactions, where the temperature inside the particle is greater than the one at the surface, leading to higher reaction rates in the core region of the catalyst.

For a first-order reaction rate, it is possible to analytically derive the dependency between the effectiveness factor and the Thiele modulus:

$$\eta = 3\frac{1}{\phi_1^2}(\phi_1 \coth \phi_1 - 1) \tag{3.27}$$

The steady-state intraparticle profiles, with related effectiveness factor values at different Thiele modulus, are given in Example 3.1.

Example 3.1 First-order reaction, intraparticle profiles

For a first-order reaction it is possible to calculate the dimensionless concentration inside the particle adopting the following analytical solution of the intraparticle mass balance equation:

$$y = \frac{C_A}{C_{A,Bulk}} = \frac{1}{x}\frac{\sinh(\phi_1 x)}{\sinh(\phi_1)}$$

As revealed, the dimensionless concentration profiles are calculated as a function of the Thiele modulus and the dimensionless radial coordinate.

For this special case, the effectiveness factor is calculated from the Thiele modulus as it follows:

$$\eta = 3\frac{1}{\phi_1^2}(\phi_1 \coth \phi_1 - 1)$$

Solve both the equations for ϕ_1= 0.1:0.1:6, plotting y versus x for ϕ_1 = 0.1,1,2,4,6 and η versus ϕ_1 for the whole range.

Matlab code for the solution of this example and results are reported as follows Figure 3.4:

```
% example 3.1
clc, clear

x = 0:1e-3:1 ;
TM = 0.1:0.1:6;
```

```
for j=1:length(TM)
 for k=1:length(x)
   y(k,j) =(1/x(k))*(sinh(TM(j).*x(k))./sinh(TM(j))) ;
  end
   eta(j) = (3/(TM(j)^2))*(TM(j)*coth(TM(j))-1) ;
end

figure(1)
subplot(1,2,1)
plot(x,y(:,find(TM==0.1)),'-r',x,y(:,find(TM==1)),'-k',x,y(:,find(TM==2)),'-b',. . .
  x,y(:,find(TM==4)),'-y',x,y(:,find(TM==6)),'-g')
xlabel('\it x\rm [-]'), ylabel('\it y\rm [-]')
legend('\phi_{1} = 0.1','\phi_{1} = 1','\phi_{1} = 2','\phi_{1} = 4','\phi_{1} = 6')

subplot(1,2,2)
semilogx(TM,eta)
xlabel('\phi_{1} [-]'), ylabel('\eta [-]')
```

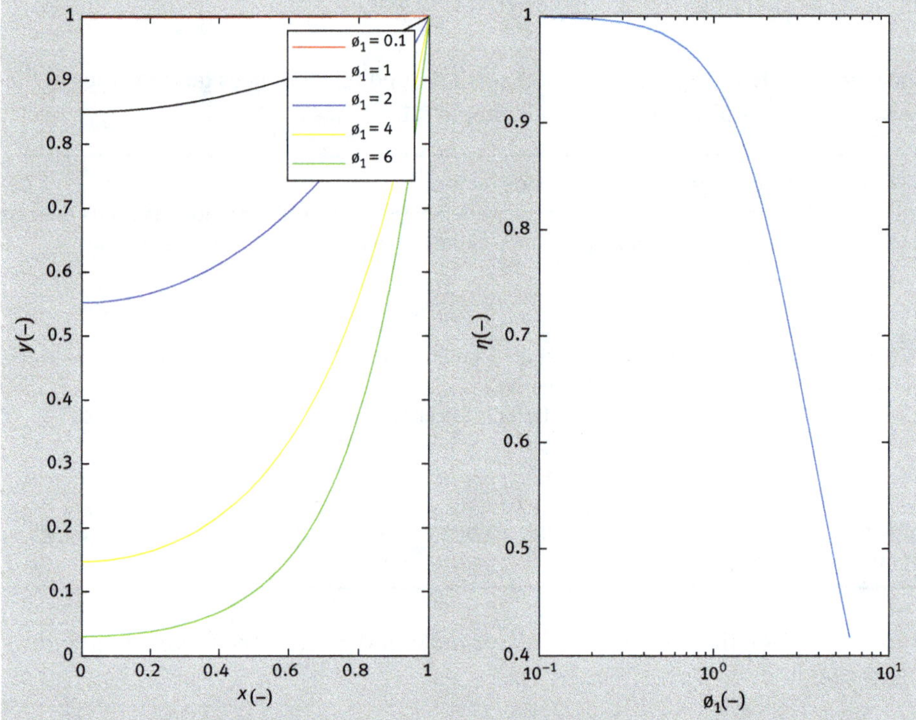

Figure 3.4: (Left) Dimensionless concentration profile as a function of the dimensionless radial coordinate of the particle. (Right) Effectiveness factor plot versus the Thiele modulus.

As revealed, higher Thiele modulus leads to higher concentration gradients inside the particle and a lower effectiveness factor.

3.2.3 Simulation strategy: *pdepe* MATLAB function

For a single-particle problem, a PDE system must be solved. In some cases, analytical solutions are reported in the literature, but they are generally valid for simple tasks. A very efficient way to solve numerically the issue with MATLAB is the use of the *pdepe* function, which allows to solve PDE systems, fixing both initial and boundary conditions. Calling the function, several input values must be given:

$$\mathrm{sol} = \mathrm{pdepe}(\mathrm{m}, \mathrm{pdefun}, \mathrm{icfun}, \mathrm{bcfun}, \mathrm{xmesh}, \mathrm{tspan}, \mathrm{opts}); \tag{3.28}$$

- *s* is the shape factor: MATLAB accepts only 0, 1 and 2 as values
- *pdefun* is a .m file where the PDEs is defined
- *bcfun* is a .m file where the boundary conditions specific of the problem are defined
- *icfun* is a .m file where the initial conditions are defined
- *xmesh* is a vector that defines the numerical grid in terms of radial coordinate where the user needs the solutions
- *tspan* is the time vector
- *opts* to define options
- *sol* is a multidimensional matrix containing the solution

The *pdepe* function requires that the PDEs are written in the following form:

$$c\left(x, t, u, \frac{\partial u}{\partial x}\right)\frac{\partial u}{\partial t} = x^{-s}\frac{\partial u}{\partial x}\left(x^s f\left(x, t, u, \frac{\partial u}{\partial x}\right)\right) + s\left(x, t, u, \frac{\partial u}{\partial x}\right) \tag{3.29}$$

where u, in our case, is either the concentration or temperature and c, f and s the coefficients to be defined. Their values strictly depend on the way how the PDE is written.

For further details, the reader is cordially invited to vision the MATLAB help.

Starting from this approach, several examples are given in this book. For instance, Example 3.2 deals with the solution of the mass balance equations for a single reaction, while Example 3.3 the simultaneous solution of mass and heat balances equations.

Example 3.2 Single particle, isothermal reactor
For a given reaction A → B, solve the intraparticle mass balance for a first-order reaction kinetics, adopting the single particle approach. The settings needed are listed in Table 3.1.

Table 3.1: Settings needed to solve the single particle modeling.

Parameter	Value	Units
D_e	1e−9	m^2/s
C_B	1e3	mol/m^3

Table 3.1 (continued)

Parameter	Value	Units
R_p	1e−3	m
t	0:20	s
k_m	1	m/s
ε	0.5	–
ρ_B	1,000	kg/m^3
k	1e−4	m^3/(kg s)

Consider the particle free of any reactants and products at $t = 0$.

Matlab code for the solution of this example and results are reported as follows Figure 3.5:

```
% example 3.2
clc,clear

%% Settings
k=1e-4;         % m3/(kg s)
De=1e-9;        % m2/s
CB=1e3;         % mol/m3
Rp=1e-3;        % m
rmesh=linspace(0,Rp,100);
tspan=linspace(0,20,101);
rhoP=1000;      % kg/m3
km=1;           % m/s
eps=0.5;        % -

%% pdpe call
m=2;

f1=@(pl,ul,pr,ur,t)bcfun3_02(pl,ul,pr,ur,t,CB,km,De);
f2=@(rp,t,u,dudx)pdefun3_02(rp,t,u,dudx,eps,k,De,rhoP);
f3=@(rp)icfun3_02(rp);
c=pdepe(m,f2,f3,f1,rmesh,tspan);
%% Figures

subplot(1,2,1)
plot(rmesh,c(2,:),'-r',rmesh,c(21,:),'-b',rmesh,c(61,:),'-k')
xlabel('\it r_{p} \rm [m]')
ylabel('\it C_{A} \rm [mol/m^{3}]')
grid on
legend('\it t\rm= 0.2s','\it t\rm= 4s','\it t\rm= 12s')
```

```
subplot(1,2,2)
surf(rmesh,tspan,c,'EdgeColor','none')
xlabel('\it r_{p} \rm [m]')
ylabel('\it t \rm [s]')
zlabel('\it C_{A} \rm [mol/m^{3}]')

function [c,f,s]=pdefun3_02(rp,t,u,dudx,eps,k,De,rhoP)
ca=u(1);
r=k*ca;      % mol/(kg s)
c(1,1)=eps;
f(1,1)=De*dudx(1);
s(1,1)=rhoP*(-r);
end

function [pl,ql,pr,qr]=bcfun3_02(pl,ul,pr,ur,t,CB,km,De)
pl(1,1)=0;
ql(1,1)=1/De;
pr(1,1)=-km*(CB-ur(1));
qr(1,1)=1;
end

function u0=icfun3_02(rp)
u0(1,1)=0;
end
```

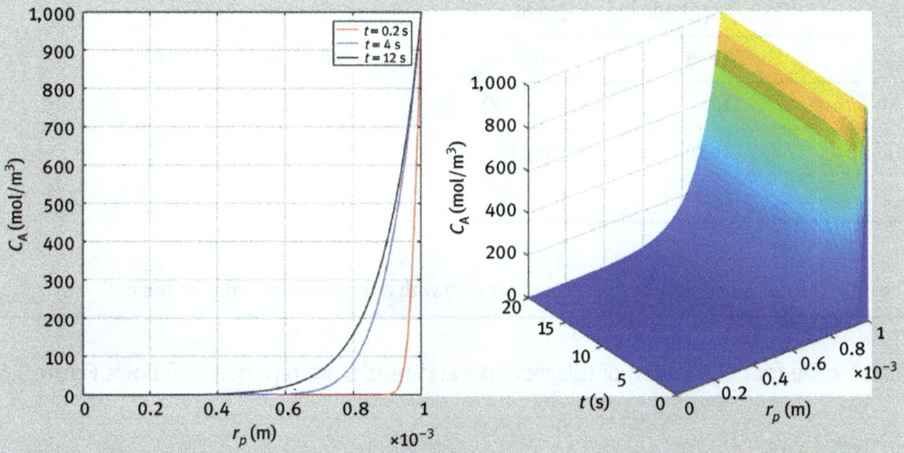

Figure 3.5: (Left) Concentration profiles as a function of the radial coordinate of the particle at different reaction times. (Right) Concentration of the reactant surface plot versus reaction time and particle radial coordinate.

The reactants diffuse inside the particle and reacts along the time as expected.

Example 3.3 Single particle, nonisothermal reactor

For a given reaction A → B, solve the intraparticle mass balance for a first order reaction kinetics, adopting the single particle approach considering the case of a nonisothermal reactor. The settings needed are listed in Table 3.2.

Table 3.2: Settings needed to solve the single particle modeling.

Parameter	Value	Units
D_e	5e−8	m^2/s
C_B	1e3	mol/m^3
T_B	340	K
R_p	1e−3	m
t	0:15	s
k_m	1	m/s
ε	0.5	–
ρ_B	1,000	kg/m^3
k_{ref}	1e−4	$m^3/(kg\ s)$
T_{ref}	340	K
Ea	120,000	J/mol
$\Delta_r H$	−10,000	J/mol
$k_{T,p}$	1e−8	W/(m K)
h	1	$W/(m^2\ K)$
$c_{p,p}$	1e3	J/(kg K)
$c_{p,A}$	4e3	J/(mol K)

Consider the particle empty of the reactant at $t = 0$ with a temperature equal to the bulk one.

Matlab code for the solution of this example and results are reported as follows Figure 3.6:

```
% example 3.3
clc,clear

%% Settings
kref=1e-4;      % m3/(kg s)
Tref=340;       % K
Ea=120000;      % J/mol
```

```
R=8.3144;       % J/(K mol)
DrH=-10000;     % J/mol

De=5e-8;        % m2/s
CB=1e3;         % mol/m3
Rp=1e-3;        % m
rmesh=linspace(0,Rp,100);
tspan=linspace(0,20,101);
rhoP=1000;      % kg/m3
km=1;           % m/s
eps=0.5;        % -

ktp=1e-8;       % W/(m K)
h=1;            % W/(m2 K)
TB=340;         % K

%% pdpe call
m=2;

f1=@(pl,ul,pr,ur,t)bcfun3_03(pl,ul,pr,ur,t,CB,km,De,TB,ktp,h);
f2=@(rp,t,u,dudx)pdefun3_03(rp,t,u,dudx,eps,kref,Ea,R,Tref,DrH,De,rhoP,ktp);
f3=@(rp)icfun3_03(rp,TB);

c=pdepe(m,f2,f3,f1,rmesh,tspan);

%% Figures
subplot(1,2,1)
plot(rmesh,c(2,:,1),'-r',rmesh,c(21,:,1),'-b',rmesh,c(61,:,1),'-k')
xlabel('\it r_{p} \rm [m]')
ylabel('\it C_{A} \rm [mol/m^{3}]')
grid on
legend('\it t\rm= 0.2s','\it t\rm= 4s','\it t\rm= 12s')

subplot(1,2,2)
plot(rmesh,c(2,:,2),'-r',rmesh,c(21,:,2),'-b',rmesh,c(61,:,2),'-k')
xlabel('\it r_{p} \rm [m]')
ylabel('\it T \rm [K]')
grid on
legend('\it t\rm= 0.2s','\it t\rm= 4s','\it t\rm= 12s')

function [c,f,s]=pdefun3_03(rp,t,u,dudx,eps,kref,Ea,R,Tref,DrH,De,rhoP,ktp)
ca=u(1);
T=u(2);

k=kref*exp(-Ea/R*(1/T-1/Tref)); % m3/(kg s)
r=k*ca;         % mol/(kg s)
cpp=1e3;        % J/(kg K)
cpa=4e3;
```

```
alfa=eps*cpa*ca+rhoP*cpp*(1-eps);

c(1,1)=eps;
f(1,1)=De*dudx(1);
s(1,1)=rhoP*(-r);

c(2,1)=alfa;
f(2,1)=ktp*dudx(2);
s(2,1)=-DrH*rhoP*r;
end

function [pl,ql,pr,qr]=bcfun3_03(pl,ul,pr,ur,t,CB,km,De,TB,ktp,h)
pl(1,1)=0;
ql(1,1)=1/De;
pr(1,1)=-km*(CB-ur(1));
qr(1,1)=1;

pl(2,1)=0;
ql(2,1)=1/ktp;
pr(2,1)=-h*(TB-ur(2));
qr(2,1)=1;

end
function u0=icfun3_03(rp,TB)
u0(1,1)=0;
u0(2,1)=TB;

end
```

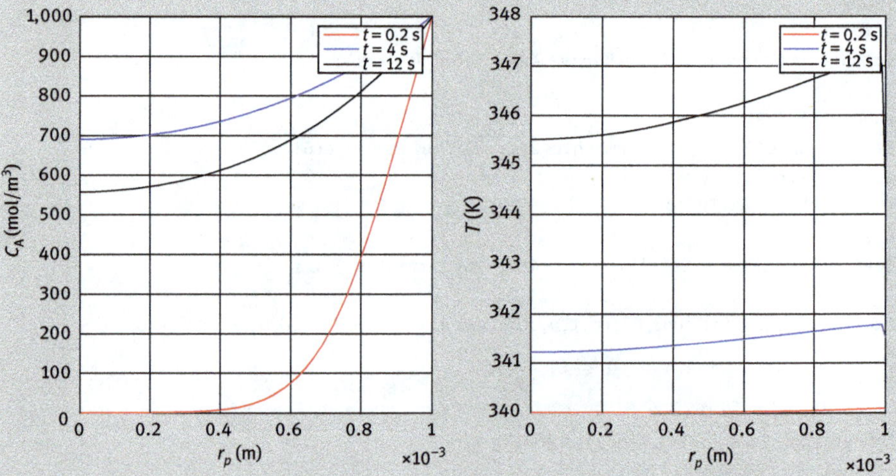

Figure 3.6: (Left) Concentration profiles as a function of the radial coordinate of the particle at different reaction times. (Right) Temperature profiles as a function of the radial coordinate of the particle at different reaction times.

The reactant diffuses and reacts. As the reaction is exothermic, a temperature increases within the particle is observed, which increases with time as the reaction proceeds. Moreover, the temperature peak is located where the concentration is the highest, and it moves to the inner part when the particle gets filled with the reactant.

3.2.4 Nonuniform active phase distributions

To maximize the performances of heterogeneous catalytic reactors, it is necessary to consider many variables, for example, catalytic particle dimension, shape and active phase distribution. Many examples reported in the literature showed that nonuniform active-phase distributions lead to the optimization of the performance of the reactor overall for reactions in series where the intermediate is the desired product [4–7]. To simulate nonuniform active-phase profiles, distribution functions need to be defined. Four main active-phase profiles are reported in the literature: (i) uniform: the active phase if homogeneously distributed in the support; (ii) eggshell: the active phase is located in the outer surface of the support; (iii) egg white: the active phase is included in a region between the outer shell and the inner core; (iv) egg yolk: the active phase is present in the inner core of the support.

Analytical functions (Ω_k) are proposed, whose codomain varies from 0 to 1. Defining b as smoothing factor (needed to modify the function steepness), a_i and a_{ij} are the coordinates of the distribution domains, it is possible to define a function for each catalyst typology:

$$\text{Egg shell} \quad 0.5 + 0.5\tanh\left(\frac{x-a_1}{b}\right) \tag{3.30}$$

$$\text{Egg yolk} \quad 0.5 - 0.5\tanh\left(\frac{x-a_2}{b}\right) \tag{3.31}$$

$$\text{Egg white} \quad 0.5\tanh\left(\frac{x-a_{32}}{b}\right) - 0.5\tanh\left(\frac{x-a_{31}}{b}\right) \tag{3.32}$$

These distributions are directly inserted into eqs. (3.12) and (3.17), multiplicator of the reaction rate:

$$\varepsilon\frac{\partial C_i}{\partial t} = x^{-s}\frac{\partial}{\partial x}\left(\frac{D_{e,i}}{R_p^s}x^s\frac{\partial C_i}{\partial x}\right) + \sum_{k=1}^{N_R}\Omega_k \nu_{ik} r_k \rho_p \tag{3.33}$$

$$\left(c_{p,p}\rho_p(1-\varepsilon) + \sum_{i=1}^{N_C} c_{p,i}C_i\varepsilon\right)\frac{\partial T}{\partial t} = x^{-s}\frac{\partial}{\partial x}\left(\frac{k_{T,p}}{R_p^s}\frac{\partial T}{\partial x}x^s\right) + \sum_{k=1}^{N_R}(-\Delta_r H_k)\Omega_k r_k \rho_p \tag{3.34}$$

In this way, in the regions of the catalyst where $\Omega_k = 1$, the generation term is active, opposite when $\Omega_k = 0$, the generation term is zero; thus, the solid material becomes inert.

Two different examples are reported, where the reader can appreciate the difference of using either eggshell or egg yolk catalyst for an in-series reaction network with a given kinetics.

Example 3.4 Single reaction, isothermal reactor, eggshell catalyst
A → B reaction is promoted by an eggshell catalyst. Solve the intraparticle mass balance for a first-order reaction kinetics, considering the case of isothermal reactor. The settings needed are listed in Table 3.3.

Table 3.3: Settings needed to solve the single particle modeling for an eggshell catalyst.

Parameter	Value	Units
D_e	5e−8	m^2/s
C_B	1e3	mol/m^3
R_p	1e−3	m
t	0:20	s
k_m	1	m/s
ε	0.5	–
ρ_B	1,000	kg/m^3
k	1e−4	$m^3/(kg\ s)$
a_1	0.6e−3	–
b	1e−5	–

Consider the particle empty of the reactant at $t = 0$.

Matlab code for the solution of this example and results are reported as follows Figure 3.7:

```
% example 3.4
clc,clear

%% Settings
a1=0.6e-3;   % Inflection point
b=1e-5;      % Smoothing factor
k=1e-4;      % m3/(kg s)
De=5e-8;     % m2/s
```

```
CB=1e3;       % mol/m3
Rp=1e-3;      % m
rmesh=linspace(0,Rp,100);
tspan=linspace(0,20,101);
rhoP=1000;    % kg/m3
km=1;         % m/s
eps=0.5;      % -

%% pdpe call
m=2;

f1=@(pl,ul,pr,ur,t)bcfun3_04(pl,ul,pr,ur,t,CB,km,De);
f2=@(rp,t,u,dudx)pdefun3_04(rp,t,u,dudx,eps,k,De,rhoP,a1,b);
f3=@(rp)icfun3_04(rp);

c=pdepe(m,f2,f3,f1,rmesh,tspan);

%% Figures
subplot(1,2,1)
plot(rmesh,c(2,:,1),'-r',rmesh,c(21,:,1),'-b',rmesh,c(61,:,1),'-k')
xlabel('\it r_{p} \rm [m]')
ylabel('\it C_{A} \rm [mol/m^{3}]')
grid on
legend('\it t\rm= 0.2s','\it t\rm= 4s','\it t\rm= 12s')

subplot(1,2,2)
surf(rmesh,tspan,c,'EdgeColor','none')
xlabel('\it r_{p} \rm [m]')
ylabel('\it t \rm [s]')
zlabel('\it C_{A} \rm [mol/m^{3}]')

function [c,f,s]=pdefun3_04(rp,t,u,dudx,eps,k,De,rhoP,a1,b)
ca=u(1);

r=k*ca;         % mol/(kg s)

omega=distr_fun3_04(rp,a1,b);

c(1,1)=eps;
f(1,1)=De*dudx(1);
s(1,1)=rhoP*(-r)*omega;

end

function fx=distr_fun3_04(rp,a1,b)
fx=0.5+0.5*tanh((rp-a1)/b) ;

end
```

```
function [pl,ql,pr,qr]=bcfun3_04(pl,ul,pr,ur,t,CB,km,De)
pl(1,1)=0;
ql(1,1)=1/De;
pr(1,1)=-km*(CB-ur(1));
qr(1,1)=1;

end

function u0=icfun3_04(rp)
u0(1,1)=0;

end
```

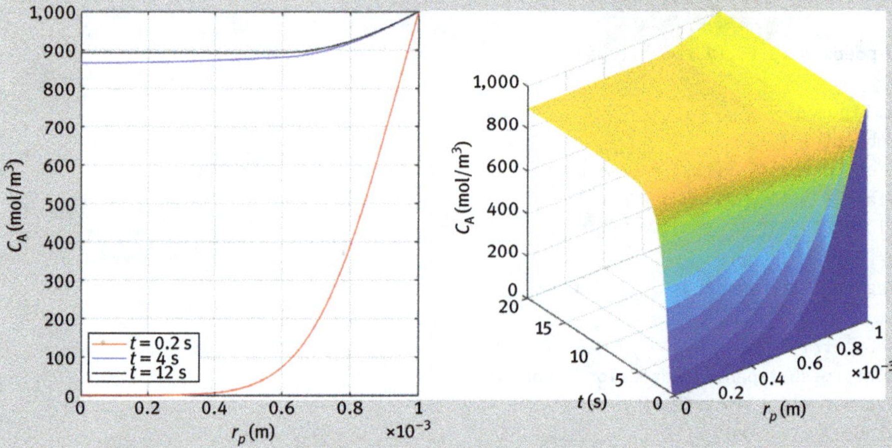

Figure 3.7: (Left) Concentration profiles as a function of the radial coordinate of the particle at different reaction times. (Right) Concentration of the reactant surface plot versus reaction time and particle radial coordinate.

As the active phase is located in the outer part of the catalyst particle, the reaction occurs, thus a concentration change is observed in that region. In the inner core, the particle is inert, the unique physical phenomena that occurs is diffusion, thus the gradient becomes less pronounced than the outer shell one.

Example 3.5 Single reaction, isothermal reactor, egg yolk catalyst

A → B reaction is promoted by an egg yolk catalyst. Solve the intraparticle mass balance for a first order reaction kinetics, considering the case of isothermal reactor. The settings needed are listed in Example 3.4, with the exception of the distribution function parameters listed in Table 3.4.

Table 3.4: Settings needed to solve the single particle modeling for an eggshell catalyst.

Parameter	Value	Units
a_2	0.6e−3	–
b	1e−5	–

Consider the particle empty of the reactant at $t = 0$.

The new distribution function and the solution of this example and results are reported as follows:

```
% example 3.5

function fx=distr_fun3_05(rp,a2,b)
fx=0.5-0.5*tanh((rp-a2)/b);

end
```

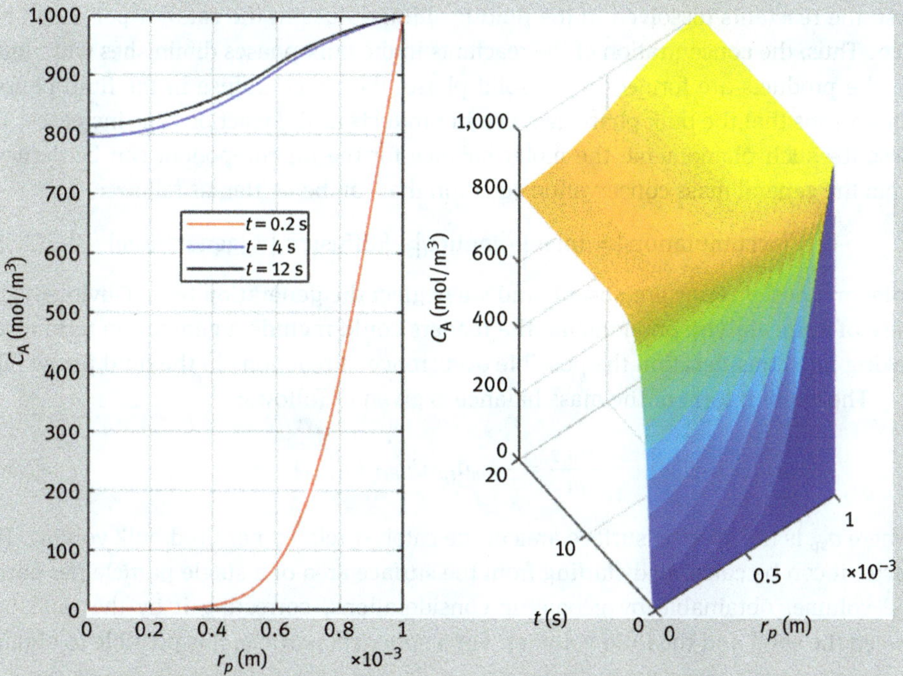

Figure 3.8: (Left) Concentration profiles as a function of the radial coordinate of the particle at different reaction times. (Right) Concentration of the reactant surface plot versus reaction time and particle radial coordinate.

The reaction occurs only in the inner part of the particle where the active phase is present. In general, lower conversions are observed compared to the eggshell catalyst case.

3.3 Fluid–solid batch reactors

In this paragraph, an attempt is reported to simulate simultaneously the mass and every balance equation for a fluid–solid system, considering both bulk fluid and intraparticle phase dynamics. This situation represents the modeling of a classical batch system when a heterogeneous catalyst promotes the reaction. The theory beyond is the same as in the previous paragraphs. What is needed is the simultaneous solution of the ODE system used to describe the fluid bulk phase (see Chapter 2) and the PDE system for the intraparticle domain.

3.3.1 Mass and energy balances derivation

From a general point of view, when a fluid phase is put in contact with a solid catalyst, the reactants dissolved in the fluid medium diffuse in the catalyst pores and reacts. Thus, the concentration of the reactants in the bulk phases diminishes with time. As the products are formed in the solid phase, they back diffuse in the fluid phase. This means that the bulk phase gets rich in products as the reaction time increases. To describe such phenomena, the molar balance for the *i*th component can be derived from the general mass conservation equation that can be written as follows:

$$[\text{accumulation}_i] = [\text{inlet}_i] - [\text{outlet}_i] + [\text{diffusion}_i] + [\text{generation}_i] \tag{3.35}$$

Inlet and outlet terms are absent, and we neglect the generation term. Obviously, in case of autocatalytic phenomena, the readers could include a generation term also, taking into consideration the possible occurrence of reactions in the fluid phase [3].

The explicit form of the mass balance is given as follows:

$$\frac{dC_{i,B}}{dt} = -k_m a_{sp}\left(C_{i,B} - C_i|_{x=1}\right) \tag{3.36}$$

where a_{sp} is the specific surface area of the catalyst related per fluid bulk volume. De facto, it can be calculated starting from the surface area of a single particle per particle volume, obtainable by geometric considerations, correcting it by the ratio between the solid and the fluid volumes. For a spherical particle, it is possible to obtain

$$a_{sp} = \frac{n_p A}{V_B} = \frac{n_p A}{n_p V}\frac{n_p V}{V_B} = \frac{A}{V}\frac{n_p V}{V_B} = \frac{4\pi R_p^2}{\frac{4}{3}\pi R_p^3}\frac{n_p V}{m_p}\frac{m_p}{V_B} = \frac{3}{R_p}\frac{\rho_B}{\rho_p} \tag{3.37}$$

where n_p is the total number of particles, A the surface of a single particle, V the volume of a single particle, V_B the volume of the fluid bulk phase, R_p the particle radius, ρ_B the bulk density of the catalyst and ρ_P the catalyst density.

In detail, the ratio between the two phases is expressed in terms of the ratio between the bulk density of the catalyst and the density of the catalyst itself, both measurable quantities.

Naturally, if the intraparticle diffusion limitations are negligible, that is, small particle size, high effective diffusivity, the mass balance equation is simpler to be solved, having the form reported below, corresponding to a pseudohomogeneous model where the catalyst is imagined to be dissolved in the fluid phase:

$$\frac{dC_{i,B}}{dt} = \sum_{k=1}^{Nr} v_{ik} r_k \rho_B \tag{3.38}$$

In an analogous way, the heat balance equation can be derived:

$$c_{p,B} \frac{m_B}{V_B} \frac{dT_B}{dt} = -h a_{sp} \left(T_B - T|_{x=1} \right) \tag{3.39}$$

where h is the heat transfer coefficient.

3.3.2 Simulation strategy: finite difference method

For a complete fluid–solid batch reactor model, it is not possible to use *pdepe* MATLAB function, as the whole system is comprised of mixed ODE and PDE equations that must be solved simultaneously and *pdepe* lacks this opportunity at least till December 2019. The only way is to discretize the partial derivatives, approximating them to finite difference methods, which means that derivatives are treated as discrete intervals:

$$\frac{\partial C_i}{\partial x} \cong \frac{\Delta C_i}{\Delta x} \tag{3.40}$$

The discretization is a tricky issue, as different methods are reported in the literature depending on the approximation. In general, it is possible to classify the following approximations, taking the first derivative calculation for each case as example:

- *Backward*: derivative evaluated using the back points

$$\frac{\partial C_i}{\partial x} \cong \frac{-C_i(x_{-1}) + C_i(x)}{\Delta x} \tag{3.41}$$

- *Forward*: derivative evaluated using the forward points

$$\frac{\partial C_i}{\partial x} \cong \frac{+C_i(x) - C_i(x_{+1})}{\Delta x} \tag{3.42}$$

- *Central*: derivative evaluated using both back and front points

$$\frac{\partial C_i}{\partial x} \cong \frac{-0.5C_i(x_{-1}) + 0C_i(x) + 0.5C_i(x_{+1})}{2\Delta x} \tag{3.43}$$

In general, the central approximation is very accurate, but it can lead to fluctuations when solving second derivatives. The backward approach is normally used when facing with flow reactor, as the feed point is always certain. When higher accuracy is needed, it is possible to increase the number of points for the calculation of the derivative. In Tables 3.5 and 3.6, the coefficients needed for the first and second derivative calculations for different accuracy degrees are reported for, respectively, central finite difference and backward/forward methods.

Table 3.5: Central finite difference method coefficients for the calculation of first and second derivatives. Point 0 is the location where the derivative is calculated.

Derivative	Accuracy	−4	−3	−2	−1	0	1	2	3	4
1	2				−1/2	0	1/2			
	4			1/12	−2/3	0	2/3	−1/12		
	6		−1/60	3/20	−3/4	0	3/4	−3/20	1/60	
	8	1/280	−4/105	1/5	−4/5	0	4/5	−1/5	4/105	−1/280
2	2				1	−2	1			
	4			−1/12	4/3	−5/2	4/3	−1/12		
	6		1/90	−3/20	3/2	−49/18	3/2	−3/20	1/90	
	8	−1/560	8/315	−1/5	8/5	−205/72	8/5	−1/5	8/315	−1/560

Table 3.6: Backward/forward finite difference method coefficients for the calculation of first and second derivatives. Point 0 is the location where the derivative is calculated.

Derivative	Accuracy	0	1	2	3	4	5	6	7
1	1	−1	1						
	2	−3/2	2	−1/2					
	3	−11/6	3	−3/2	1/3				
	4	−25/12	4	−3	4/3	−1/4			
	5	−137/60	5	−5	10/3	−5/4	1/5		
	6	−49/20	6	−15/2	20/3	−15/4	6/5	−1/6	

Table 3.6 (continued)

2	1	1	−2	1					
	2	2	−5	4	−1				
	3	35/12	−26/3	19/2	−14/3	11/12			
	4	15/4	−77/6	107/6	−13	61/12	−5/6		
	5	203/45	−87/5	117/4	−254/9	33/2	−27/5	137/180	
	6	469/90	−223/10	879/20	−949/18	41	−201/10	1019/180	−7/10

It is evident that the correct implementation of the finite difference method is not trivial, and it cannot be generalized, as it is case dependent.

3.4 Three-phase batch reactors

Facing with multiphase systems means solving simultaneously the ODE and PDE systems that derives from the derivation of the mass and heat balance equations specific of the case study. The scheme reported in Figure 3.9 depicts a general case where either a gas–liquid–solid or liquid–liquid–solid system is present [8].

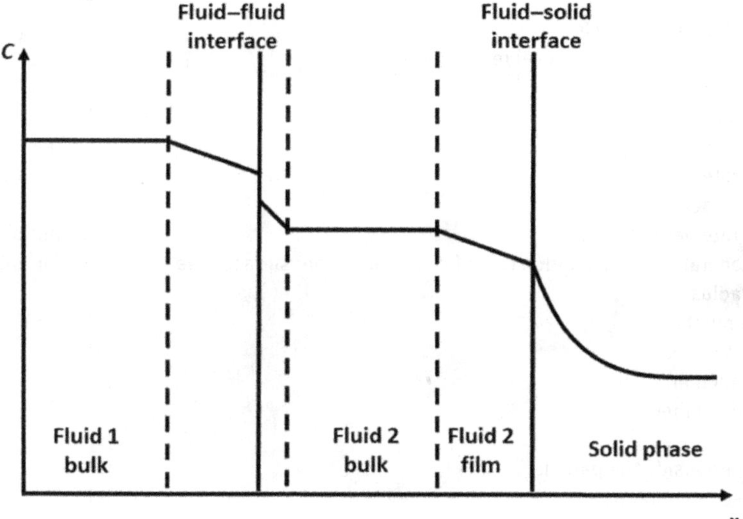

Figure 3.9: Mass transfer in fluid–fluid–solid system.

Even if it looks complicated, the general idea is to start from the concepts already illustrated in Chapter 2 and the ones in this chapter. The authors are sure that using the given examples, the reader will be able to reproduce such more sophisticated model.

List of symbols

A	Catalyst surface	[m^2]
a_i, a_{ij}	Coordinates for the active phase distribution function	[–]
a_{sp}	Catalyst specific surface per fluid bulk volume	[m^2/m^3]
b	Smoothing factor	[–]
D_i	Molecular diffusivity	[m^2/s]
$D_{e,i}$	Effective diffusivity	[m^2/s]
d_p	Particle diameter	[m]
C_i	Concentration of component i in the solid phase	[mol/m^3]
$C_{i,B}$	Concentration of component i in the fluid bulk phase	[mol/m^3]
$c_{p,i}$	Fluid specific heat	[J/(mol K)]
$c_{p,B}$	Fluid specific heat	[J/(kg K)]
$c_{p,p}$	Catalyst specific heat	[J/(kg K)]
h	Fluid–solid heat transfer coefficient	[W/(m^2 K)]
J_i	Mass transfer flow of component i	[mol/(m^2 s)]
k	Kinetic constant of ith chemical reaction	[*]
k_m	Fluid–solid mass transfer coefficient of component i	[m/s]
$k_{T,p}$	Catalyst heat conductivity	[W/(m K)]
L	Diffusion length	[m]
m	Mass	[kg]
n_p	Number of catalyst particles	[–]
N_R	Number of independent chemical reactions	[–]
Nu	Nusselt number	[–]
Pr	Prandtl number	[–]
r	Reaction rate	[mol/(kg s)]
r_k	Reaction rate of reaction k	[mol/(kg s)]
r_p	Particle radial coordinate	[m]
r_s	Reaction rate per surface area	[mol/(m^2 s)]
$R_{i,s}$	Overall generation/consumption rate of component i per surface area	[mol/(m^2 s)]
R_p	Particle radius	[m]
Re	Reynolds number	[–]
s	Shape factor	[–]
Sc	Schmidt number	[–]
Sh	Sherwood number	[–]
t	Time	[s]
T	Temperature inside the particle	[K]
T_B	Bulk phase temperature	[K]
u	Generic variable	[–]
U	Fluid velocity	[m/s]
V	Catalyst volume	[m^3]
V_B	Fluid bulk phase volume	[m^3]
x	Dimensionless coordinate	[–]

Greek letters

δ	Fluid film thickness	[m]
$\Delta_r H_k$	Reaction enthalpy	[J/mol]
ε	Particle porosity	[–]
ρ_B	Catalyst bulk density	[kg/m^3]
ρ_p	Density of the catalyst	[kg/m^3]
σ	Constriction factor	[–]
τ	Particle tortuosity	[–]
v_i	Stoichiometric coefficient of component *i*	[–]
$v_{i,k}$	Stoichiometric coefficient of component *i* in reaction *k*	[–]
v	Fluid viscosity	[m^2/s]
μ	Fluid viscosity	[Pa s]
ϕ	Thiele modulus	[–]
η	Effectiveness factor	[–]
Ω_k	distribution function	[–]

*Units of kinetic constants depend on the expression of the reaction rate.

References

[1] H.S. Fogler. Elements of Chemical Reaction Engineering (5th edition). Prentice Hall: 2016.

[2] E. Santacesaria, R. Tesser. The Chemical Reactor from Laboratory to Industrial Plant. Springer: 2008.

[3] V. Russo, V. Hrobar, P. Mäki-Arvela, K. Eränen, F. Sandelin, M. Di Serio, T. Salmi. Kinetics and Modelling of Levulinic Acid Esterification in Batch and Continuous Reactors. Topics in Catalysis 2018, 1–10.

[4] V. Russo, L. Mastroianni, R. Tesser, T. Salmi, M. Di Serio. Intraparticle Modeling of Non-Uniform Active Phase Distribution Catalyst. ChemEngineering 2020, 4, 24.

[5] M. Morbidelli, A. Varma. On shape normalization for non-uniformly active catalyst pellets. Chemical Engineering Science 1983, 38, 297–305.

[6] C. Perego, P. Villa. Catalyst preparation methods. Catalysis Today 1997, 34, 281–305.

[7] N.M. Deraz. The importance of catalyst preparation. Journal of Industrial and Environmental Chemistry 2018, 2, 16–18.

[8] M. Di Serio, V. Russo, E. Santacesaria, R. Tesser, R. Turco, R. Vitiello. Liquid-liquid-solid model for the epoxidation of soybean oil catalysed by Amberlyst-16. Industrial & Engineering Chemistry Research 2017, 56(45), 12963–12971.

Chapter 4
Ideal single-phase continuous reactors for homogeneous catalysis

4.1 Introduction

Continuous reactors are the most frequently used devices, in industry, when high productivity is required, for example, in bulk chemical syntheses or for chemical productions at commodity scale. Depending on the flow behavior, ideal continuous reactors that are treated in this chapter can be classified as continuous tubular plug flow reactors (PFR) or continuous stirred tank reactors (CSTR). Even if only homogeneous catalysis is considered in this chapter, both the presence of a single phase (liquid or gaseous) and two phases (gas–liquid or liquid–liquid) will be described. As reference on the topic, some classical textbooks can be examined [1–4].

4.2 Continuous plug flow reactor

The development of both material and energy balance for a continuous tubular reactor is based, for an ideal reactor, on the following assumptions:

- The reactor is in stationary conditions so that concentration and temperature are constant in a certain position in the reactor. The case of nonstationary tubular reactors will be treated subsequently in a dedicated chapter (Chapter 6).
- The motion of the fluid in the reactor is like the motion of piston in an internal combustion engine. This assumption involves that temperature and concentration in a certain section of the reactor are constant and no radial profiles are developed.

These two simplified assumptions lead to a compact form of material and energy balance represented by a system of ordinary differential equations (ODE) that must be integrated along the reactor volume. For a mixture of N components, N ODEs are enough for the description of an isothermal reactor, while additional thermal balance is required for nonisothermal system. In a schematized way, this kind of reactors can be represented as in Figure 4.1 in which a simple isothermal reactor is reported as case "a," while a nonisothermal jacketed reactor is referred to as case "b." In this last configuration, a thermostatting fluid is circulated in the jacket with the aim of controlling the temperature of the system.

https://doi.org/10.1515/9783110632927-004

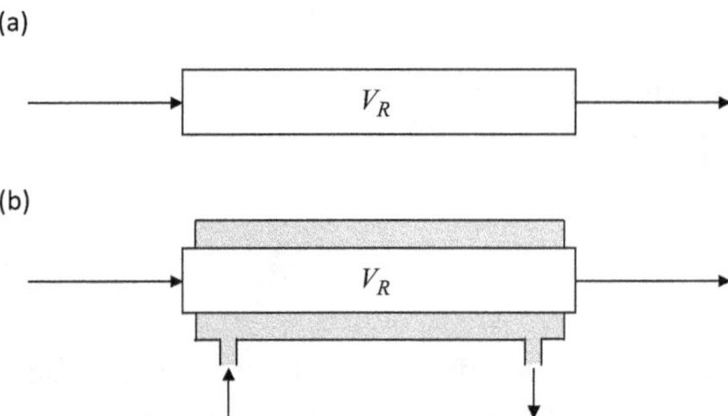

Figure 4.1: General scheme for PFR reactors. Case a: isothermal reactor; case b: nonisothermal jacketed reactor.

4.2.1 Mass balance, isothermal PFR reactor

The mass balance for a continuous PFR can be developed, related to a generic component A, by referring to the scheme reported in Figure 4.2. The reactor has a simple cylindrical geometry with constant cross-sectional area *A* and length *L* [1].

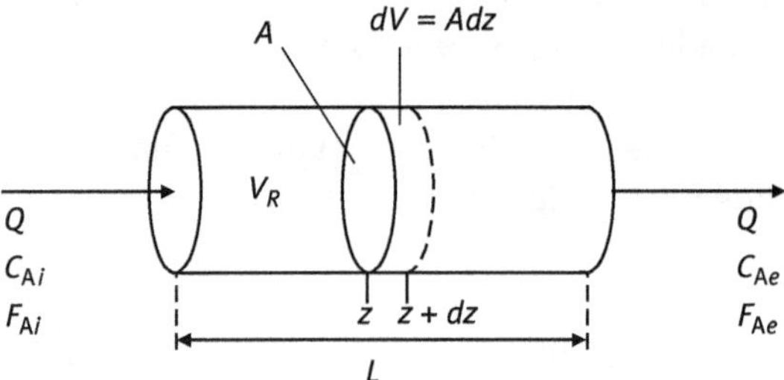

Figure 4.2: Scheme for mass balance.

Consider, as a first approach, a system that could be assumed in liquid phase in which a single homogeneous reaction takes place. In this case, the mole balance for the *i*th component can be derived from the general mass conservation equation that can be written as:

$$[\text{accumulation}_i] = [\text{inlet}_i] - [\text{outlet}_i] + [\text{generation}_i] \quad (4.1)$$

This general conservation equation can be simplified as the reactor is assumed in stationary state and the accumulation term is null:

$$[\text{inlet}_i] + [\text{generation}_i] = [\text{outlet}_i] \quad (4.2)$$

Considering the differential reactor volume dV, between z and $z + dz$, we can write eq. (4.2) in terms of molar flow rate:

$$F_i + v_i r dV = F_i + dF_i \tag{4.3}$$

or

$$dF_i = v_i r dV \tag{4.4}$$

and finally,

$$\frac{dF_i}{dV} = v_i r \tag{4.5}$$

In eqs. (4.3)–(4.5), F_i is the molar flow rate of component i, v_i is the stoichiometric coefficient of component i, r is the reaction rate and V is the volume coordinate of the reactor. For Nc components, a set of Nc ODEs like eq. (4.5) must be solved simultaneously by numerical integration from $V = 0$ to $V = V_R$. The initial conditions for such integration are represented by the feed to the reactor F_{Ai}.

Equation (4.5) can be written in a more general form for a system in which Nr chemical reactions occur as follows:

$$\frac{dF_i}{dV} = \sum_{j=1}^{Nr} v_{i,j} r_j \tag{4.6}$$

Obviously, in this case, a rectangular matrix of stoichiometric coefficients ($Nc \times Nr$) must be defined in order to solve the system of ODEs. An equivalent form of eq. (4.5) that could be useful in some cases, in terms of concentration instead of molar flow rates (referred to a single reaction), is

$$u \frac{dC_i}{dV} = v_i r \tag{4.7}$$

In relation (4.7), u is the linear velocity of the fluid entering into the reactor. The linear velocity of the fluid can be evaluated, in the case of incompressible fluid, as the ratio:

$$u = \frac{Q}{A} \tag{4.8}$$

where Q is the volumetric flow rate that feed the reactor and A is the cross-sectional area of the reactor itself.

From a practical point of view, the integration of the ODE system (4.5)–(4.6) or (4.7) requires the knowledge of the kinetic expression for r_i and the related kinetic parameters. The most common form for a simple kinetic expression of a chemical reaction is represented, for a system in liquid phase, by the following equation:

$$r_i = k_i \prod_{j=1}^{Nc_i} C_j^{\lambda_j} \tag{4.9}$$

where k_i is the kinetic constant, Nc_i is the number of components involved in the reaction i, C_j is the concentration of the component and λ_j is reaction order with respect to the component j. Obviously, instead of concentrations in eq. (4.9), mole numbers can be directly substituted instead of concentrations giving place to the following expression:

$$r_i = k_i \prod_{j=1}^{Nc_i} \left(\frac{n_j}{V_R}\right)^{\lambda_j} \tag{4.10}$$

A possibility that occurs frequently for continuous tubular reactors is that of gas-phase reactions. In this case, concentrations are replaced by partial pressure in the derivation of reaction rate expression:

$$r_i = k_i \prod_{j=1}^{Nc_i} p_j^{\lambda_j} = k_i \prod_{j=1}^{Nc_i} (Py_j)^{\lambda_j} \tag{4.11}$$

where p_j is the partial pressure of jth component, P is the total pressure and y_j is the mole fraction in gaseous phase related to component j. When information about the eventual nonideality of gaseous mixture is available, fugacities should be used instead of partial pressures and a suitable equation of state can be adopted, in this case, for calculations. For a constant pressure system, the mathematical model is completely defined when a constitutive equation is coupled with ODEs (4.5) or (4.6) as follows:

$$y_i = \frac{F_i}{\sum_{j=1}^{NC} F_j} \tag{4.12}$$

If the total pressure of the system is not constant, for example, when temperature profile is operative in the reactor or when pressure drop is considered with an equation like that of Ergun [x05] (valid for packed beds), a further algebraic or differential equation must be coupled with eq. (4.5) or (4.6). The same approach should be used when the total number of moles in the gaseous mixture changes as a consequence of the chemical reactions network. A simpler way to introduce a pressure variation along the tubular reactor is represented by the following relation:

$$P = P_0 \frac{F}{F_0} \frac{T}{T_0} \tag{4.13}$$

In this expression, the subscript "0" refers to the reactor inlet condition, while the variable without subscript is related to a certain position in the reactor. F is the overall molar flow rate defined by the expression:

$$F = \sum_{j=1}^{N_C} F_j \tag{4.14}$$

Example 4.1 Multiple reactions, constant volumetric flow rate, liquid phase, constant pressure, isothermal PFR reactor

In a PFR, a set of four multiple reactions, with a complex network, occur involving five components identified as A, B, C, D and E. The reaction scheme is as follows:

$$\begin{aligned} &A + B \longleftrightarrow 2C + D \\ &A + C \rightarrow E \\ &D \rightarrow 2B \end{aligned}$$

The first reaction is reversible, so forward and reverse reactions must be accounted for. On the basis of this scheme, the following stoichiometric matrix can be arranged in which the stoichiometric coefficients of all the components in all the reactions are included.

Table 4.1: Stoichiometric matrix of the reaction network in the Example 4.1.

Component	Reaction 1	Reaction 2	Reaction 3	Reaction 4
A	−1	+1	−1	0
B	−1	+1	0	+2
C	2	−2	−1	0
D	1	−1	0	−1
E	0	0	1	0

For a complete definition of the problem, the expressions of the liquid-phase reaction rates are necessary, together with the values of the corresponding kinetic constants:

$$\begin{aligned} r_1 &= k_1 C_A C_B \\ r_2 &= k_2 C_C^2 C_D \\ r_3 &= k_3 C_A \\ r_4 &= k_4 C_D \end{aligned}$$

The kinetic constants related to the four reactions have the values: 1*e*−5, 2*e*−6, 3*e*−5 and 1*e*−5; the reactor volume is 6 $m^3$and the volumetric feed flow rate is $Q = 0.003\ m^3/s$. The molar feed flow rate of the five single components A to E are, respectively, the following: 1.5, 1.2, 0, 0 and 0.

Develop a Matlab code that can solve the differential material balance equations and plot the concentration profiles of all the components along the reactor volume. In a compact form, the material balance for the tubular reactor can be written as:

$$\frac{dF_i}{dt} = \sum_{k=1}^{Nr} v_{i,k} r_k, \quad i = \text{A, B, C, D, E}, \; N_R = \text{fourreactions}$$

Matlab code for the solution of this example and results are as follows Figure 4.3:

```
% example 4.1

clc,clear
global k Q ni Nc Nr

Nc=5;                                % number of components (-)
Vr=6;                                % reactor volume (m3)
Nr=4;                                % number of reactions (-)
Q=0.003;  % volumetric feed flow rate (m3/s)
k=[0.01 0.002 0.03 0.01]/1000;  % kinetic constants
ni=[-1 +1 -1 0;                      % matrix of stoichiometric coefficients
   -1 +1 0 +2;                       % row -> component
   +2 -2 -1 0;                       % column -> reaction
   +1 -1 0 -1
   0 0 +1 0];

f0=[1.5 1.2 0 0 0];                  % feed molar flow rates (mol/s)
vspan=0:Vr/1000:Vr;                  % volume range for integration (m3)

[tx,fx]=ode15s(@ode_ex4_01,vspan,f0);
%% plot
subplot(1,2,1)
plot(tx,fx(:,1),tx,fx(:,2),tx,fx(:,3))
grid
xlabel('Volume (m3)')
ylabel('Molar flow rate (mol/s)')
legend('A','B','C')

subplot(1,2,2)
plot(tx,fx(:,4),tx,fx(:,5))
grid
xlabel('Volume (m3)')
ylabel('Molar flow rate (mol/s)')
legend('D','E')

function [df] = ode_ex4_01(v,f)
global k Q ni Nc Nr
```

```
c=f/Q;
r(1)=k(1)*c(1)*c(2);
r(2)=k(2)*c(3)^2*c(4);
r(3)=k(3)*c(1)*c(3);
r(4)=k(4)*c(4);

for j=1:Nc
  df(j)=0;
  for jj=1:Nr
    df(j)=df(j)+r(jj)*ni(j,jj);
  end
end

df=df';
end
```

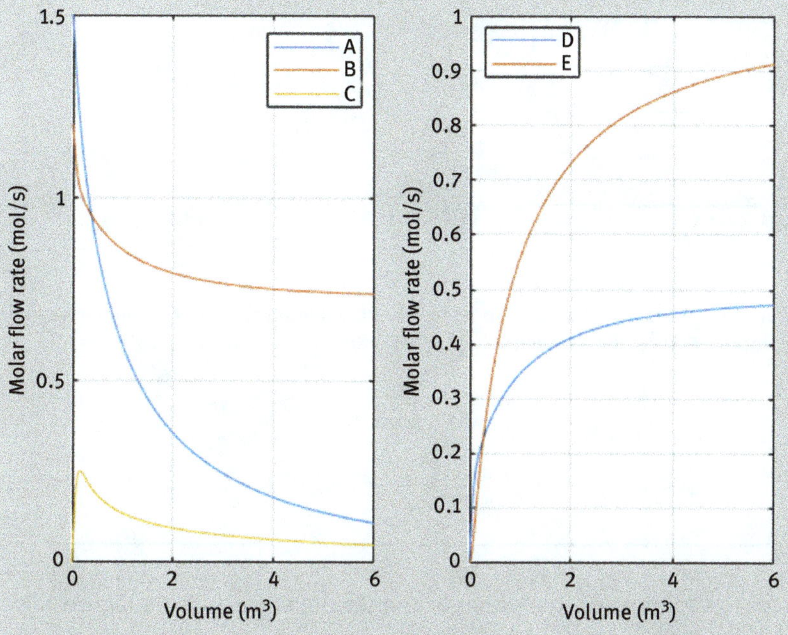

Figure 4.3: (Left) Molar flow rates of the components A, B and C versus reactor volume; (right) molar flow rates of components D and E versus reactor volume.

Example 4.2 Single reaction, constant volumetric flow rate, gas phase, variable pressure, isothermal PFR reactor

In a tubular PFR reactor ($Vr = 1\ \text{m}^3$), a single gas-phase reaction occurs with the following stoichiometry:

$$A + B \rightarrow C + 2D$$

This reaction is characterized by a second-order kinetics with a reaction rate expression represented by the following relation in which the kinetic constant is $k = 0.5\ \text{mol/(m}^3\ \text{s atm}^2)$:

$$r = kP_A P_B$$

The molar flow rates of each component fed to the reactor are summarized in the following table:

Table 4.2: Feed flow rates of the various components, example 4.02.

Component	Feed flow rates (mol/s)
A	1
B	1.2
C	0
D	0

The inlet pressure of the reactor is 8 atm. To describe the volume profiles of components molar flow rate and of total pressure, the following coupled differential equations must be solved:

$$\frac{dF_A}{dV_R} = -r \quad \frac{dF_B}{dV_R} = -r \quad \frac{dF_C}{dV_R} = +r \quad \frac{dF_D}{dV_R} = +2r$$

$$P_{tot} = P_{tot}^0 \frac{F_{tot}}{F_{tot}^0} \quad y_i = \frac{F_i}{\sum_{i=1}^{N_C} F_i} = \frac{F_i}{F_{tot}} \quad p_i = P_{tot} y_i \quad i = A, B, C, D$$

Matlab code for the solution of this example and results are as follows Figure 4.4:

```
% example 4.2
clc,clear
global Vr ni k P0 F0
Vr=1;                  % reactor volume (m3)
ni=[-1 -1 1 2];        % stoich. coefficients (-)
k=0.5;                 % kinetic constant (mol/(s m3 atm2)
P0=8;                  % inlet total pressure (atm)
F0=[1 1.2 0 0];        % feed flow rates (mol/s)
vspan=0:Vr/1000:Vr;    % volume range for integration (m3)
```

```
[vx,Fx]=ode45(@ode_ex4_02,vspan,F0);
for j=1:length(vx)
  Px(j)=P0*sum(Fx(j,:))/sum(F0); % total pressure (atm)
end
%% plot
subplot(1,2,1)
plot(vx,Fx(:,1),vx,Fx(:,2),vx,Fx(:,3),vx,Fx(:,4))
grid
xlabel('Volume (m3)')
ylabel('Molar flow rates (mol/s)')
legend('A','B','C','D')
subplot(1,2,2)
plot(vx,Px)
grid
xlabel('Volume (m3)')
ylabel('Total pressure (atm)')

function [dF] = ode_ex4_02(t,F)
global ni k P0 F0

Ftot0=sum(F0);
Ftot=sum(F);          % total molar flow rate (mol/s)
y=F./Ftot;            % mole fractions (-)
P=P0*Ftot/Ftot0;      % total pressure (atm)
p=P*y;                % component partial pressure (atm)
r=k*p(1)*p(2);        % reaction rate (mol/(m3 s))
dF=ni*r;              % mass balances (mol/(s m3))

dF=dF';
end
```

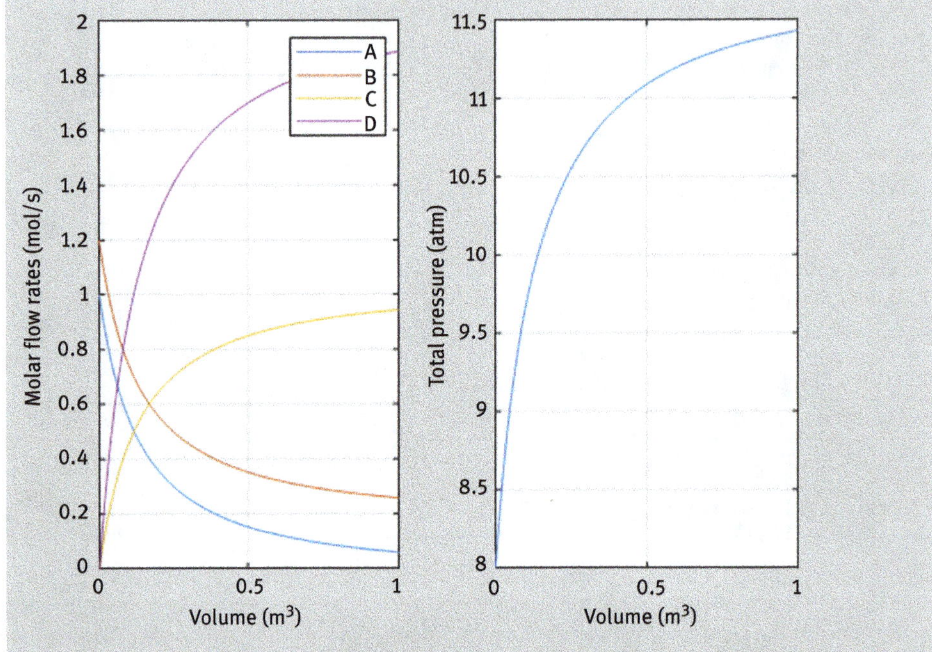

Figure 4.4: (Left) Molar flow rates of the components A, B, C and D versus reactor volume; (right) total pressure versus reactor volume.

4.2.2 Energy balance

In the derivation of material balances related to continuous PFR reactor, presented in Section 4.2.1, the temperature was assumed constant in the whole reactor in analogy to what was done for batch reactor (BR) described in the Chapter 2. Reaction temperature affects mainly kinetics, and kinetic constants are consequently fixed in the equations of Section 4.2.1. Another effect linked to temperature variation along the reactor consists in influencing the total pressure as illustrated by relation (4.13).

When we focus on energy balance, as expected, we must couple to mass balances a further conservation equation related to heat. In the present section, we will derive such relation that, depending on the modalities of heat exchange characteristics of the tubular continuous reactor, is able to describe the thermal behavior of the PFR.

The general form of energy conservation equation is as follows:

$$[\text{heat accumulated}] = [\text{inlet heat}] - [\text{outlet heat}] + [\text{heat generated}] + [\text{heat exchanged}] \tag{4.15}$$

As we have assumed that the reactor is in stationary conditions, the energy accumulation term is null, and the energy balance is reduced to the following relation:

$$[\text{outlet heat}] - [\text{inlet heat}] = [\text{heat generated}] + [\text{heat exchanged}] \tag{4.16}$$

The relation (4.16) can be expressed in terms of temperature derivative as

$$Q\,\rho C_{\mathrm{pm}}\frac{dT}{dV} = Q_{\mathrm{Re}} - Q_E \tag{4.17}$$

In this expression, Q is the volumetric feed flow rate, ρ is the average mixture density, C_{pm} is the average specific heat of the reacting mixture, Q_R is the heat released or absorbed by all the reactions and Q_E is the heat exchanged between the reactor and the surrounding. This last term accounts for the heat exchange modality of the reactor; if Q_E is set to 0, reactor works in adiabatic condition while, if Q_E is very high, isothermal conditions are operative. In all the intermediate cases between isothermal and adiabatic, a suitable expression of heat exchange should be used as constitutive equation. The thermal behavior of the reactor can also be schematized by considering the values of the heat transfer coefficient:

- $U = 0$ adiabatic reactor, no heat is transferred from/to the reactor
- $U = +\infty$ isothermal reactor, all the heat generated is removed by the thermal control system
- $U =$ value the reactor shows an intermediate behavior between isothermal and adiabatic one

A widely used option consists in expressing heat exchanged as a function of thermal gradient and an overall heat transfer coefficient U, as in the following expression:

$$Q_E = UA_S(T - T_J) \tag{4.18}$$

where U is the overall heat transfer coefficient, A_S is the specific surface area, T is the temperature in a specific reactor location, calculated by differential equation (4.17) and T_J is the heating/cooling jacket temperature. The evaluation of the overall heat transfer coefficient, U, is crucial and can be accomplished by using suitable literature correlations [6, 7]. If the adopted heating/cooling system is characterized by a high capacity, as usually occurs, the jacket temperature is assumed constant, otherwise a heat balance on jacket fluid must be coupled with the reactor energy balance.

For what concerns the heat released or absorbed by the chemical reactions, exothermic or endothermic, can be evaluated as the sum of all the reaction rates, each of them multiplied by the correspondent heat of reaction, as in the following equation:

$$Q_{\mathrm{Re}} = -\sum_{j=1}^{N_R} \Delta H_j r_j \tag{4.19}$$

The summation in relation (4.19) is extended to all the occurring reactions and the initial minus sign accounts for the convention that assumes a negative value for the enthalpy of exothermic reaction and, on the contrary, a positive value for endothermic ones. Equation (4.19) is the link between energy and material balances as in this relation appear, implicitly in r_j, the concentrations or the partial pressure of all the chemical compounds.

A particular consideration should be made for the specific area A_S. This parameter, for cylindrical reactors, assumes the value of

$$A_S = 2/R \tag{4.20}$$

This means that for improving heat transfer efficiency, when a great amount of heat must be exchanged, low diameter tubes are the better choice.

For a practical purpose, in order to have an explicit temperature derivative, eq. (4.17) can be algebraically manipulated to obtain an operative expression like the following:

$$\frac{dT}{dV} = \frac{Q_{\text{Re}} - Q_E}{Q\rho C_{\text{pm}}} \tag{4.21}$$

or more explicitly,

$$\frac{dT}{dV} = \frac{-\sum_{j=1}^{N_R} \Delta H_j r_j - UA_S(T - T_J)}{Q\,\rho C_{\text{pm}}} \tag{4.22}$$

A possible refinement of the thermal balance model (eq. (4.22)) consists in using, for the specific heat and for the density of the reacting mixture, a function of temperature and composition instead of an average value.

The coupling of mass and heat balance (and eventually a further ODE for pressure profile) produce a system of $Nc + 1$ initial values ODEs that can be solved numerically to calculate, simultaneously, time profiles for molar flow rate component and reactor temperature.

As already stated in Chapter 2, the temperature in each position of the reactor affects in a strong way also the reaction rates through the temperature dependence of the kinetic constant. The well-known Arrhenius equation accounts for this functional dependency:

$$k_i = k_{i\,\text{ref}} \exp\left[\frac{E_{Ai}}{R_g}\left(\frac{1}{T_{\text{ref}}} - \frac{1}{T}\right)\right] \tag{4.23}$$

where $k_{i\,\text{ref}}$ is the kinetic constant referred to a reference temperature T_{ref}, R_g is the gas constant, T is the absolute temperature and E_{Ai} is the activation energy for the reaction i.

Example 4.3 Single reaction, constant volume, liquid phase, nonisothermal PFR reactor
In a constant volume PFR reactor, a single reaction occurs in liquid phase, with the following stoichiometry:

$$A + B \rightarrow C + D$$

This reaction is characterized by a second-order kinetics and its parameters are reported in Table 4.3:

Table 4.3: Rate expression and kinetic parameters for Example 4.3.

Rate expression	k_{tref}	***Ea* (J/mol)**	***ΔHr* (J/mol)**
$r = kC_AC_B$	4,000 (m^3/(mol s))	25,000	−50,000

The feed of the reactor is summarized in Table 4.4:

Table 4.4: Molar fees flow rated for Example 4.3.

Component	**Molar flow rates (mol/s)**
A	100
B	90
C	5
D	2

Other useful data regarding reactor characteristics and physicochemical properties of the mixture are summarized in Table 4.5.

Table 4.5: Physicochemical properties and reactor settings for Example 4.3.

Property	**Value**	**Units**
Reactor volume	3	(m^3)
Overall heat transfer coefficient	1,200	(J/(s m^2 K))
Reactor radius	0.03	(m)
Average density of mixture	1,100	(kg/m^3)
Average specific heat	4,000	(J/(kg K))
Heating/cooling fluid temperature	300	(K)

Table 4.5 (continued)

Property	Value	Units
Inlet reactor temperature	310	(K)
Reference temperature	323	(K)
Volumetric feed flow rate	0.001	(m^3/s)

To describe the volume profiles of both components' concentration and reaction temperature, the following equation must be solved simultaneously:

$$\frac{dF_A}{dV} = -r \quad \frac{dF_B}{dV} = -r \quad \frac{dF_C}{dV} = +r \quad \frac{dF_D}{dV} = +r$$

$$\frac{dT}{dV} = \frac{Q_r - Q_E}{Q\rho_m C_{pm}}$$

$$C_i = \frac{F_i}{Q} \quad Q_E = UA(T - T_J) \quad Q_r = -r\Delta H_r$$

Matlab code for the solution of this example and results are as follows Figure 4.5:

```
% example 4.3

clc,clear
global Qvol U DH RR Ts rom cpm Ea Tref kref
U = 1200;              % overall heat transfer coefficient (J/(s m2 K))
DH = -50000;           % heat of reactions (J/mol)
rom = 1100;            % average mixture density (kg/m3)
cpm = 4000;            % average liquid heat capacity (J/(kg K))
RR = 0.03;             % reactor radius (m)
Ts = 300;              % temperature of jacket fluid (K)
Ea = 25000;            % activation energy (J/mol)
Tref = 323;            % reference temperature (K)
kref = 4000;           % kinetic constant (m3/(mol s))
Qvol = 0.001;          % volumetric flow rate (m3/s)
Vr = 3;                % reactor volume (m3)
n0=[100 90 5 2];       % initial moles (mol)
t0=310;                % initial temperature (K)
y0=[n0 t0];            % initial conditions
vspan=0:Vr/300:Vr;     % volume interval for integration (m3)
[t,y]=ode15s(@ode_ex4_03,vspan,y0);
```

```
%% plots
subplot(1,2,1)
plot(t,y(:,1:4))
grid
xlabel('Volume (m3)')
ylabel('Molar flow rate (mol/s)')
legend('A','B','C','D')
subplot(1,2,2)
plot(t,y(:,5))
grid
xlabel('Volume (m3)')
ylabel('Temperature (K)')

function [dy] = ode_ex4_03(t,y)
global Qvol U DH RR Ts rom cpm Ea Tref kref

f=y(1:4);        % molar flow rates
T=y(5);          % temperature
c=f*Qvol;        % concentrations
R=1.987*4.189; % gas constant

k1=kref*exp((Ea/R)*(1/Tref-1/T));
r1=k1*c(1)*c(2);

dy(1)=-r1;
dy(2)=-r1;
dy(3)=+r1;
dy(4)=+r1;

A=2/RR;
Qs=U*A*(T-Ts);
Qr=-r1*DH;

dy(5)=(Qr-Qs)/(Qvol*rom*cpm);

dy=dy';
end
```

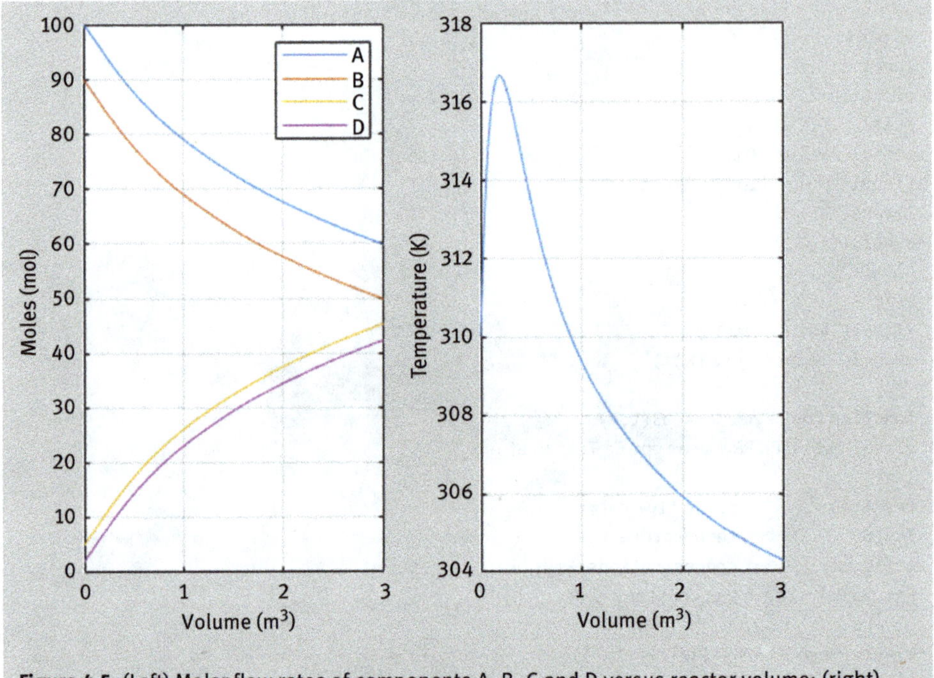

Figure 4.5: (Left) Molar flow rates of components A, B, C and D versus reactor volume; (right) reactor temperature versus reactor volume.

4.3 Two-phase plug-flow reactors

As mentioned in Chapter 2 referring to BRs, also in PFRs multiple phases present in this chapter, the simultaneous presence of gas and liquid or two liquid phases will be considered for continuous reactors. For gas–liquid systems, we make the assumption that one or more chemical reactions take place only in the liquid phase while, for liquid–liquid system, the reactions can, in principle, occur in both liquid phases.

4.3.1 Gas–liquid two-phase continuous PFR

The theory of mass transfer across contacting phases can still be adopted in continuous flow reactors. The difference between multiphase BRs and PFRs consists in the concept that in PFR, mass transfer is variable along the reactor volume while for BR it is variable in time.

If a plug flow tubular reactor is fed with both gas and liquid feeds, very likely the gas phase is dispersed as bubbles in the liquid one assumed as continuous (see Figure 4.6(a)). For modeling purpose, we can refer to the scheme reported in Figure 4.6(b) in which the overall volume of the reactor is divided in two sections: one is occupied by gas phase and the other one by liquid phase. By adopting the film resistance for mass transfer theory, and neglecting the gas-side gradient, the flux of ith component transferred from gas to liquid is

$$J_i = \beta_i \left(\frac{P_i}{H_i} - C_i \right) \tag{4.24}$$

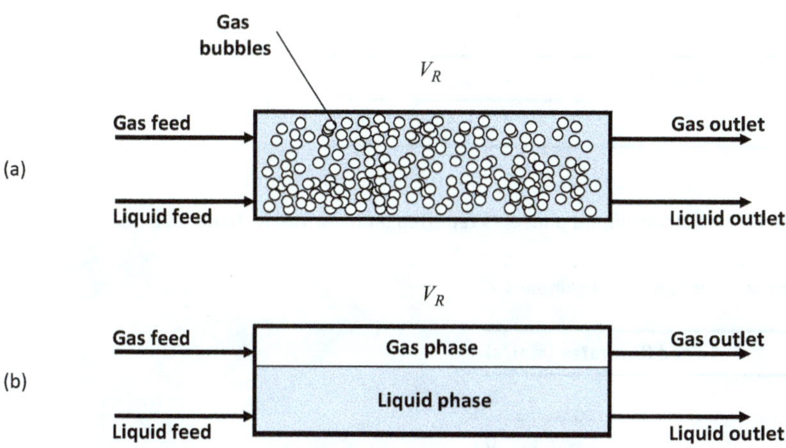

Figure 4.6: Scheme for gas–liquid tubular reactor.

This term must be included in the mass balance for a component that is partitioned between the two phases and, for that component, two separate mass balances should be written, one for each phase. Considering the PFR, the differential mass balance, modified for accounting for the simultaneous presence of two phases, is represented by the following relation:

$$\begin{cases} \frac{dF_i}{dV_k} = \nu_i \, r \text{ for components not affected by mass transfer} \\ \frac{dF_i}{dV_k} = \nu_i r + J_i \text{ for components affectedby mass transfer} \end{cases} \tag{4.25}$$

In eq. (4.25), the occurrence of a single reaction is postulated, but the balance can be easily extended to the case of multiple simultaneous reactions.

Example 4.4 PFR reactor, gas–liquid single reaction, constant volume, constant pressure, isothermal reactor

A gas–liquid isothermal PFR reactor is operated with a constant pressure in vapor phase with a single chemical reaction described by the following stoichiometry:

$$A + B \rightarrow C$$

The pressure in the gas space in the reactor is maintained constant by means of an automatic control system equipped with a backpressure regulator valve on the outlet gaseous line. Reactant A is fed into the reactor in the liquid state while B is fed as gas. This last reactant is gradually transferred to the liquid phase in which the reaction occurs along reactor volume. The kinetic of the reaction is represented by a second-order expression with a constant as in Table 4.6:

Table 4.6: Rate expression and kinetic parameters for Example 4.4.

Reaction	Rate expression	k
1	$r = kC_A C_B$	$1.2e{-}5$ ($m^3/(mol\ s)$)

The initial charge in the reactor (liquid phase) is reported summarized in Table 4.7:

Table 4.7: Molar fees flow rated for Example 4.4.

Component	Feed flow rates (mol/s)
A	2
B	0
C	0

Other useful data regarding reactor characteristics and physicochemical properties of the mixture are summarized in Table 4.8.

Table 4.8: Physicochemical properties and reactor settings for Example 4.4.

Property	Value	Units
Reactor volume	30	(m^3)
Reactor temperature	360	(K)
Constant pressure of B in gas phase	3	(atm)
Henry's constant	20	($mol/(m^3\ atm)$)
Volumetric mass transfer coefficient	0.01	(s^{-1})
Liquid feed flow rate	0.01	(m^3)
Gas feed flow rate	0.1	(m^3)

Part 1

In two subplots (one for liquid phase and one for gas phase), describe the volume profiles of the molar flow rates of the three components. Assuming as negligible the gas-side mass transfer resistance, the following equations must be solved simultaneously:

$$\frac{dF_A^L}{dV_r} = -r \quad \frac{dF_B^L}{dV_r} = -r + J_B \quad \frac{dF_C^L}{dV_r} = +r \quad \frac{dF_B^G}{dV_r} = -J_B$$

$$J_B = \beta(C_B^* - C_B) \quad C_B^* = K_H P_B$$

Part 2

Perform a parametric study by solving the same problem defined in part 1 but with various values assigned to β as in the subsequent list: 0.1; 0.01; 0.001; 0.0005; 0.0001. Report, in a third subplot, the trend of molar flow rate of component A in liquid phase for the various values of mass transfer coefficients.

Matlab code for the solution of this example and results are as follows Figure 4.7:

```
% example 4.4
clc,clear
global k Vr beta pB0
global QL QG Kh

%% part 1
Vr=30;                  % overall reactor volume (m3)
k=1.2e-5;               % kinetic constant (m3/(mol s))
beta=0.01;              % G-L mass transfer coefficient (s)
T=360;                  % temperature (K)
Kh=20;                  % Henry's constant (mol/(m3 atm))
pB0=3;                  % initial pressure of B (atm)
Rgas=0.08205/1000;      % gas constant (atm m3/(mol K))
QL=1e-2;                % liquid volum. feed flow rate (m3/s)
QG=0.1;                 % gas volum. feed flow rate (m3/s)

FA10=2;                 % liquid molar feed flow rate of A (mol/s)
FB10=0;                 % liquid molar feed flow rate of B (mol/s)
FC10=0;                 % liquid molar feed flow rate of C (mol/s)
FBg0=pB0*QG/(Rgas*T); % moles of B in gas (mol)

F0=[FA10 FB10 FC10 FBg0];
vspan=0:Vr/1000:Vr;

[vx,Fx]=ode15s(@ode_ex4_04,vspan,F0);

%% part 2 - parametric study of beta
betax=[0.1 0.01 0.001 0.0005 0.0001];
for j=1:length(betax)
```

```
  beta=betax(j);
  [vx,Fxx]=ode15s(@ode_ex4_04,vspan,F0);
  Fa(:,j)=Fxx(:,1);
end

%% plot
subplot(1,3,1)
plot(vx,Fx(:,1),vx,Fx(:,2),vx,Fx(:,3))
grid
xlabel('Volume (m3)')
ylabel('Molar flow rates (mol/s)')
legend('FAl','FBl','FCl')

subplot(1,3,2)
plot(vx,Fx(:,4))
grid
xlabel('Volume (m3)')
ylabel('Molar flow rates of B in gas (mol/s)')
legend('FBg')

subplot(1,3,3)
plot(vx,Fa)
grid
xlabel('Volume (m3)')
ylabel('Molar flow rate of A in liquid phase (mol/s)')
legend('beta=0.1','beta=0.01','beta=0.001','beta=0.0005','beta=0.0001')

function [dF] = ode_ex4_04(t,F)
global k Vr beta pB0
global QL QG Kh

FAl=F(1);
FBl=F(2);
FCl=F(3);
FBg=F(4);
VL=Vr*QL/(QL+QG);
VG=Vr*QG/(QL+QG);

cA=FAl/QL;
cB=FBl/QL;
cC=FCl/QL;
pB=pB0;
r=k*cA*cB;
```

```
cBs=Kh*pB;
J=beta*(cBs-cB);
dF(1)=-r;
dF(2)=-r+J;
dF(3)=+r;
dF(4)=-J;

dF=dF';
end
```

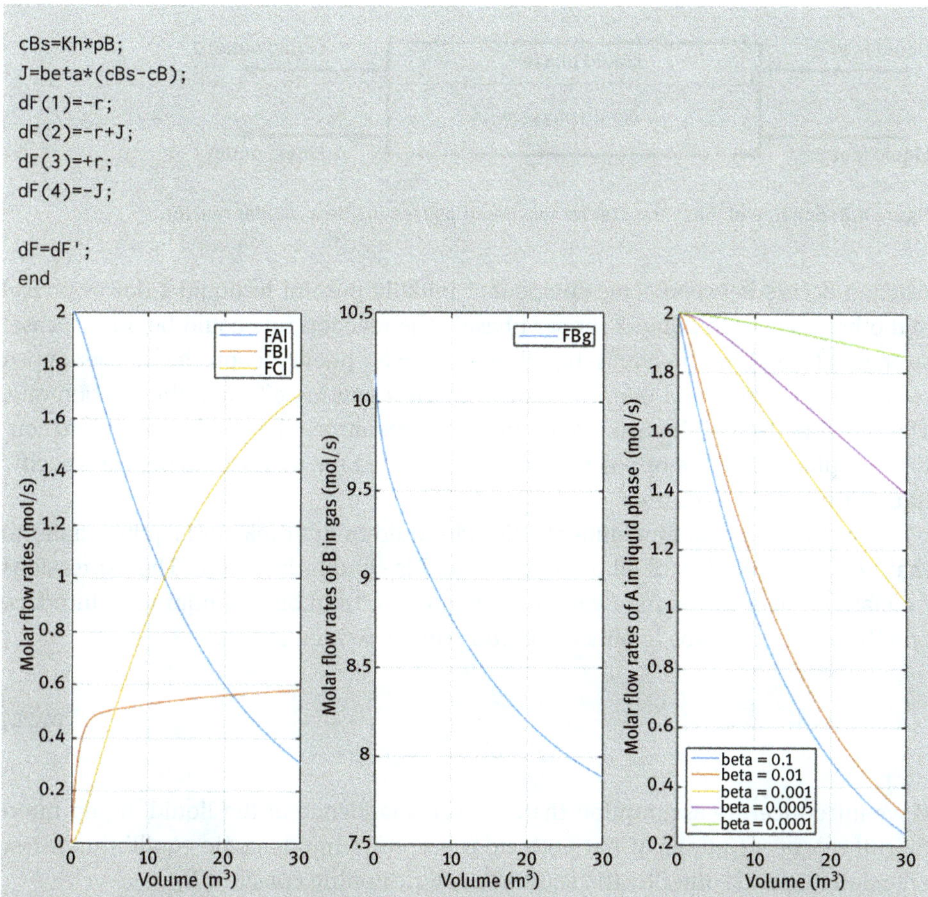

Figure 4.7: (Left) Molar flow rates of the components A, B and C in liquid phase versus reactor volume; (center) molar flow rate of B in gas phase versus reactor volume; (right) molar flow rate of components A in liquid phase for different values of mass transfer coefficient versus reactor volume.

4.3.2 Liquid–liquid two-phase tubular reactors

When two immiscible liquids are fed simultaneously to a continuous tubular reactor, the mass transfer occurs between two liquid phases and, consequently, some components are partitioned between the two contacting phases. The scheme of this particular situation is represented schematically in Figure 4.8. The real situation is more complex as, typically, one of the liquid phases is dispersed into the other one forming, in some cases, an emulsion. This is particularly relevant if a chemical

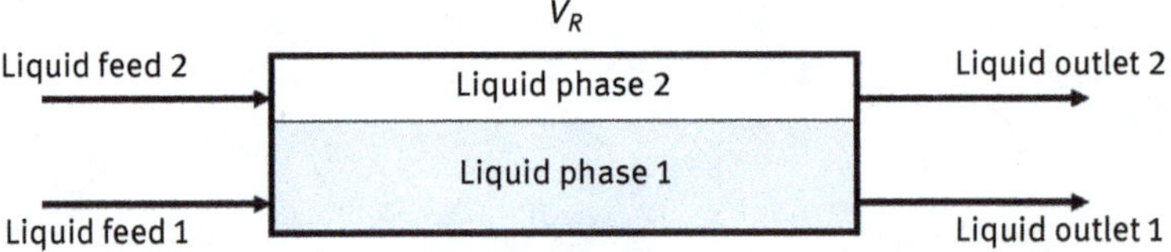

Figure 4.8: Scheme of mass transfer for two liquid phases inside a tubular reactor.

reaction occurs between some component initially present in liquid 1 (lower phase) and other present in liquid 2 (upper phase). The reaction rate could be, in this case, strongly affected by the liquid–liquid mass transfer phenomenon that is overlapped to chemical reaction and that, commonly, exerts a limiting effect on the reaction rate. The actual pattern of the two liquid phases, for example, phase 1 dispersed into continuous phase 2 or the opposite, can be accounted for by introducing the specific liquid–liquid interface area.

The mathematical modeling of a liquid–liquid tubular reactor is quite similar to that related to BR, described in Chapter 2 and is summarized here. The expressions for mass transfer flows, for a generic component i, from bulk of liquid 1 to interface and then from interface to bulk of liquid 2, can be written as

$$\begin{aligned} J_1^i &= \beta_1^i (C_1^i - C_1^{i*}) \\ J_2^i &= \beta_2^i (C_2^{i*} - C_2^i) \end{aligned} \tag{4.26}$$

If we introduce the assumption that, in correspondence of the liquid–liquid interface, the concentrations of i at both sides are in thermodynamic equilibrium, they are related to each other by the equilibrium partitioning constant $H_L{}^i$:

$$H_L^i = \frac{C_1^{i*}}{C_2^{i*}} \tag{4.27}$$

or

$$C_1^{i*} = H_L^i C_2^{i*} \tag{4.28}$$

As the concentrations at the separation interface are not easily accessible, we can assume a stationary state in correspondence of this interface (zero accumulation of partitioned components) and solve for the unknown concentration. The steady state at the interface is expressed by the relation:

$$J_1^i V_1 = J_2^i V_2 \tag{4.29}$$

The use of eq. (4.29) involves the knowledge of the amount of volume, in the reactor, occupied by liquid phases 1 and 2 and usually, in stationary conditions, we can replace volumes ratio with volumetric flow rates (Q_1 and Q_2) ratio as follows:

$$\frac{V_1}{V_2} = \frac{Q_1}{Q_2} \tag{4.30}$$

By substituting relations (4.26) and (4.30) into eq. (4.29) and rearranging algebraically, we obtain:

$$C_2^{i*} = \frac{Q_1\beta_1^i C_1^i + Q_2\beta_2^i C_2^i}{Q_1\beta_1^i H_L^i + Q_2\beta_2^i} \tag{4.31}$$

In eq. (4.31), the interface concentration of i in liquid 2 is expressed as a function only of known quantities. The liquid–liquid interface contacting area is contained in the overall mass transfer coefficient β as in the following expression:

$$\beta_k^i = k_L^i a_k \tag{4.32}$$

where $k_L{}^i$ is the liquid–liquid mass transfer coefficient of the component i and a_k is the specific interface area of the phase k.

In the derivation of differential mass balance equations, we assume, generally, that chemical reactions occur in both liquid phases and we can write, accordingly:

Liquid phase 1

$$\frac{dF_i^1}{dV} = \sum_{j=1}^{N_{r1}} \nu_{j,i} r_j^1 - J_i^1 \tag{4.33}$$

Liquid phase 2

$$\frac{dF_i^2}{dV} = \sum_{j=1}^{N_{r2}} \nu_{j,i} r_j^2 - J_i^2 \tag{4.34}$$

In relations (4.33) and (4.34), a liquid–liquid mass transfer flow, J, is present for each component. In the case of a component that is present only in one particular phase, the corresponding flux can be set to 0 or, alternatively, the partition constant H (see eq. (4.27)) must be chosen in a way that the corresponding component is completely absent in one phase.

The realistic description of a biphasic system, instead of considering it as a pseudohomogeneous one, is particularly important when the reaction network includes some catalytic species that themselves can be partitioned between liquid phases in contact, strongly affecting the reaction rates.

Example 4.5 PFR reactor, liquid–liquid system, multiple reactions, constant feed flow rate, isothermal reactor

Two liquid phases are fed to a liquid–liquid PFR reactor: aqueous phase and organic phase. The PFR reactor is operated with a constant flow rate of both liquid phases and in isothermal conditions. Two chemical reactions occur in the two liquid phases as follows:

$$A + B \rightarrow C \quad \text{aqueousphase}$$
$$B + D \rightarrow E \quad \text{organicphase}$$

According to this reaction scheme, the component B is the only substance that is partitioned between the two contacting liquid phases. The kinetics laws of the reactions are both represented by a second-order expression as in Table 4.9:

Table 4.9: Rate expression and kinetic parameters for Example 4.5.

Reaction	Rate expression	k
1 (aq.)	$r_1 = kC_AC_B$	8.0e–4 (m^3/(mol s))
2 (org.)	$r_2 = kC_BC_D$	2.0e–3 (m^3/(mol s))

The feed to the reactor, respectively, for aqueous and organic phases, is summarized in Table 4.10:

Table 4.10: Molar fees flow rated for Example 4.5.

Component	Aq. phase Molar feed flow rates (mol/s)	Org. phase Molar feed flow rates (mol/s)
A	10	10
B	10	10
C	0	0
D	0	0
E	0	0

Other useful data regarding reactor characteristics and physicochemical properties of the mixture are summarized in Table 4.11.

Table 4.11: Physicochemical properties and reactor settings for Example 4.5.

Property	Value	Units
Reactor volume	0.30	(m^3)
Mass transfer coeff. for B, aq. phase	0.6	(s^{-1})
Mass transfer coeff. for B, org. phase	0.8	(s^{-1})
Liq.–liq. partition coefficient for B	0.3	(–)
Liquid feed flow rate (aqueous phase)	0.02	(m^3/s)
Liquid feed flow rate (organic phase)	0.05	(m^3/s)

The mass balance for components in aqueous phase can be written as

$$\frac{dF_i^{aq}}{dV} = \sum_{j=1}^{N_{raq}} v_{j,i} r_j^{aq} - J_i^{aq} \quad i = A, B, C$$

$$\frac{dF_i^{org}}{dV} = \sum_{j=1}^{N_{rorg}} v_{j,i} r_j^{org} - J_i^{org} \quad i = B, D, E$$

The mass transfer flow J is defined only for the components that are partitioned between the two liquid phases B in the case of our calculation.

Solve the material balances reported earlier for the five components involved in the overall reactive system and produce two plots in which the molar flow rates are reported as a function of the reactor volume, respectively, for the aqueous phase and the organic phase.

In a third subplot, demonstrate that the overall net material balance for component B is always respected (closed) along the reactor volume.

Matlab code for the solution of this example and results are as follows Figure 4.9:

```
% example 4.5
clc,clear
global Vr k1 k2 Qa Qo
global betaa betao HB Va Vo

Vr    = 0.3;    % overall reactor volume (m3)
k1    = 8e-4;   % kinetic constant reaction 1 (m3/(mol s))
k2    = 2e-3;   % kinetic constant reaction 2 (m3/(mol s))
Qa    = 0.02;   %feed flow rate of aqueous phase (m3/s)
Qo    = 0.05;   %feed flow rate of organic phase (m3/s)
betaa = 0.6;    % mass transfer coeff. - aq. side (1/min)
betao = 0.8;    % mass transfer coeff. - org. side (1/min)
HB = 0.3;       % partition coefficient comp. B A/O (-)
Va=Vr*Qa/(Qa+Qo);
Vo=Vr*Qo/(Qa+Qo);
```

```
f0A=[10 10 0 0 0];
f0O=[ 0 0 0 12 0];
f0=[f0A f0O];
vspan=0:Vr/500:Vr;
options=odeset('RelTol',1e-13,'AbsTol',1e-13);
[v,fx]=ode15s(@ode_ex4_05,vspan,f0,options);

%% plots
subplot(2,2,1)
plot(v,fx(:,1),v,fx(:,2),v,fx(:,3))
grid
title('Aqueous phase')
legend('A','B','C')
xlabel('Reactor volume (m3)')
ylabel('Molar flow rates (mol/s)')

subplot(2,2,2)
plot(v,fx(:,7),v,fx(:,9),v,fx(:,10))
grid
title('Organic phase')
legend('B','D','E')
ylabel('Molar flow rates (mol/s)')
xlabel('Reactor volume (m3)')

bil=fx(:,2)+fx(:,3)+fx(:,7)+fx(:,10)-f0A(2);
subplot(2,2,3)
plot(v,bil)
grid
title('Overall B material balance')
xlabel('Reactor volume (m3)')
ylabel('Molar flow rates (mol/s)')

for j=1:length(v)
  [dfx,CBx]=ode_ex4_05(v(j),fx(j,:));
  CBall(j,:)=CBx;
end
subplot(2,2,4)
plot(v,CBall)
grid
title('Concentrations of B')
legend('CBa','CBaS','CBoS','CBo')
xlabel('Reactor volume (m3)')
ylabel('Concentration (mol/m3)')

function [df, CBx]=ode_ex4_05(v,f)
global Vr k1 k2 Qa Qo
global betaa betao HB Va Vo
  Fa=f(1:5);
```

```
  Fo=f(6:10);
  Ca=Fa./Qa;
  Co=Fo./Qo;

  CAa=Ca(1);
  CBa=Ca(2);
  CCa=Ca(3);
  CDa=Ca(4);
  CEa=Ca(5);

  CAo=Co(1);
  CBo=Co(2);
  CCo=Co(3);
  CDo=Co(4);
  CEo=Co(5);

  r1=k1*CAa*CBa;
  r2=k2*CDo*CBo;

  CBoS=(betaa*Va*CBa + betao*Vo*CBo) / . . .
    (betao*Vo + betaa*Va*HB);
  CBaS=HB*CBoS;
  JBa=betaa*(CBa-CBaS);
  JBo=betao*(CBoS-CBo);

%% material balance aqueous phase
  df(1) = -r1;
  df(2) = -r1 - JBa;
  df(3) = +r1;
  df(4) = 0;
  df(5) = 0;

%% material balance organic phase
  df(6) = 0;
  df(7) = -r2 + JBa; %**Vo/Va;
  df(8) = 0;
  df(9) = -r2;
  df(10)= +r2;

%% array of B concentrations
  CBx(1)=CBa;
  CBx(2)=CBaS;
  CBx(3)=CBoS;
  CBx(4)=CBo;

%% transposition of derivative arrays
  df=df';
end
```

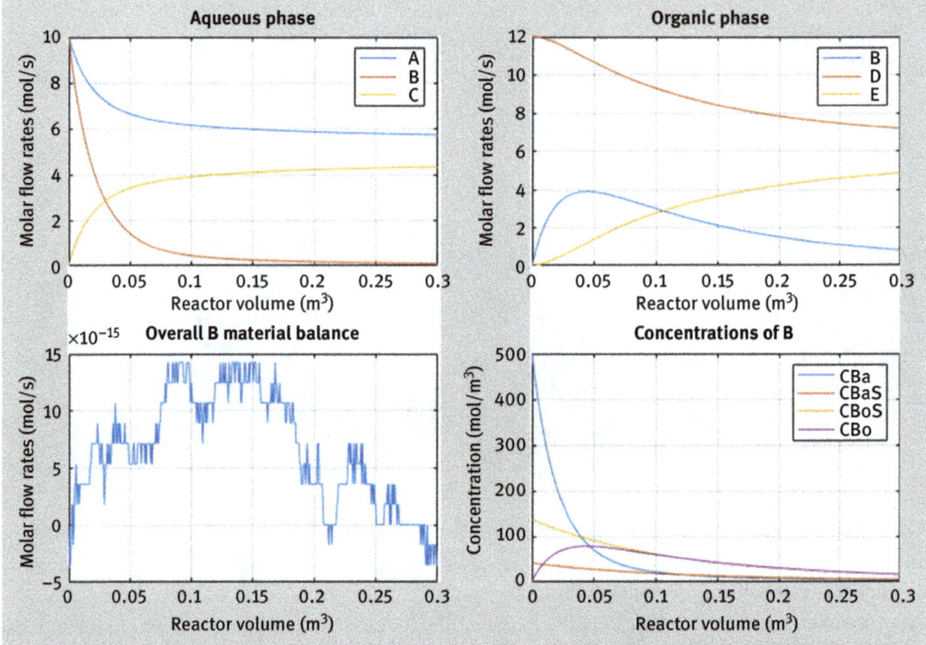

Figure 4.9: (Up-left) Molar flow rates of the components A, B and C in aqueous phase versus reactor volume; (up-right) molar flow rates of the components A, B and C in organic phase versus reactor volume; (down-left) overall material balance of B expressed as flow rate versus reactor volume; (down-right) concentration of B in both bulk phases and at the interface organic–aqueous.

4.4 Continuous stirred tank reactor

The other wide class of continuous reactors is represented by tank reactors equipped with mechanical stirring device or recirculated in a loop in order to have homogeneous concentration and temperature distributions in the whole volume.

The development of mass and energy balance for this kind of reactor can be based on the scheme reported in Figure 4.10. The fundamental assumptions for describing this kind of reactor are:

- The concentration of each component is the same in all the reaction volume, for example, spatially homogeneous.
- The temperature is the same in all the reaction volume.
- The concentrations and temperature of outlet stream are the same as the mixture inside the reactor.

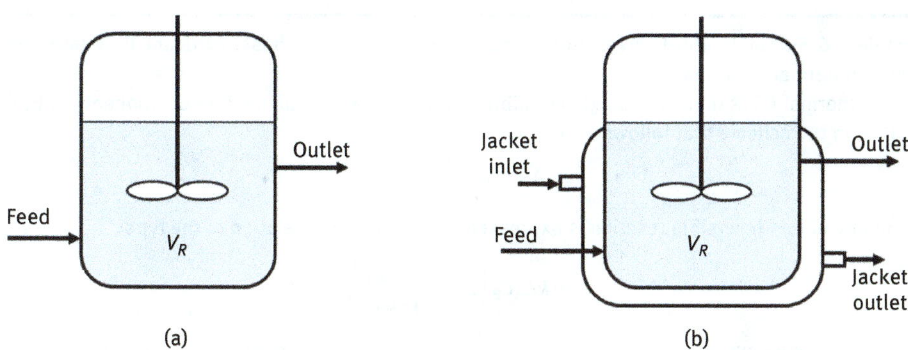

Figure 4.10: Scheme of CSTR reactor: (a) reactor without heat transfer and (b) reactor equipped with jacket for heat transfer.

4.4.1 Mass balance for a CSTR reactor

For an isothermal reactor, depicted in the scheme of Figure 4.10A, the dynamic mass balance for a generic component i, involved in Nr simultaneous chemical reactions, can be written as in the following relation:

$$\frac{dC_i}{dt} = \frac{Q_{\text{IN}} C_{i,\text{IN}}}{V_r} - \frac{Q_{\text{OUT}} C_i}{V_r} + \sum_{j=1}^{N_r} \nu_{j,i} r_j \tag{4.35}$$

In this equation, the accumulation term results from the difference between inlet and outlet and generation. In the case of constant volumetric flow rate between inlet and outlet streams, and constant reaction volume, relation (4.35) becomes

$$\frac{dC_i}{dt} = \frac{Q}{V_r}(C_{i,\text{IN}} - C_i) + \sum_{j=1}^{N_r} \nu_{j,i} r_j \tag{4.36}$$

Finally, referring to the same reactor, a steady-state mass balance can be obtained by setting the accumulation term at the left-hand side of (4.36) to zero:

$$\frac{Q}{V_r}(C_{i,\text{IN}} - C_i) + \sum_{j=1}^{N_r} \nu_{j,i} r_j = 0 \tag{4.37}$$

If the system is composed by Nc components, a set of coupled ODEs, like (4.35) or (4.36), or a system of coupled nonlinear algebraic equations (NLAEs), like (4.37), must be solved.

Example 4.6 Single equilibrium reaction, constant volume, liquid phase, constant flow rate, dynamic isothermal CSTR reactor

In an isothermal CSTR reactor, a single equilibrium reaction occurs among four components A, B, C and D as in the scheme that follows:

$$A + B \longleftrightarrow C + D$$

The kinetic of this reversible reaction is expressed by second-order relation of the type:

$$r = kC_AC_B\left[1 - \frac{1}{K_e}\frac{C_CC_D}{C_AC_B}\right]$$

The kinetic and equilibrium constants have, respectively, the values: 9 $m^3/(mol\ s)$ and 82; the reactor volume is 6 m^3 and the volumetric feed flow rate is $Q = 0.3\ m^3/s$. Inlet concentrations of the four components are 1.5, 1.2, 0 and 0.1 mol/m^3.

Part 1

Develop a Matlab code that can solve the system of nonlinear equations representing mass balances related to the four components involved in the reaction:

$$\frac{Q}{V_r}(C_{i,\mathrm{IN}} - C_i) + \sum_{j=1}^{N_r} v_{j,i}r_j = 0 \quad i = A, B, C, D$$

Part 2

The stationary solution obtained in part 1 can be used as starting values for simulating dynamic behavior of the reactor. Develop a Matlab code that can solve the ODE system represented by the following relation:

$$\frac{dC_i}{dt} = \frac{Q}{V_r}(C_{i,\mathrm{IN}} - C_i) + \sum_{j=1}^{N_r} v_{j,i}r_j \quad i = A, B, C, D$$

Integrate the system of ODE with initial conditions obtained in part 1 for a time range of 200 s. Simulate a step change in inlet concentration assuming that after 50 s the inlet concentrations are doubled.

Matlab code for the solution of this example and results are as follows Figure 4.11:

```
% example 4.6
clc,clear
global Vr Q k Ke cin Nc ni
global cinstep
Nc=4;        % number of components (-)
Nr=1;        % number of reactions (-)
Q=0.3;       % volumetric feed flow rate (m3/s)
Vr=6;        % reactor volume (m3)
Ke=82;       % equilibrium constant (-)
k=9; % kinetic constant (m3/(mol s))
```

```
ni=[-1 -1 +1 +1];        % matrix of stoichiometric coefficients (-)
cin=[1.5 1.2 0 0.1 ];    % inlet concentrations (mol/s)
cinstep=[3 2.4 0 0.2]; % new inlet concentrations (mol/s)
%% find stationary solution
cs0=cin;
cs=fsolve('nle_ex4_06',cs0)
%% dynamics
tspan=0:0.01:200;     % time range for integration (s)
c0=cs;
[tx,cx]=ode15s(@ode_ex4_06,tspan,c0);

%% plot
plot(tx,cx)
grid
xlabel('Time (s)')
ylabel('Reactor concentration (mol/m3)')
legend('A','B','C','D')
function [dc] = ode_ex4_06(t,c)
global Vr Q k Ke cin Nc ni
global cinstep
if t<50
   cxin=cin;
else
   cxin=cinstep;
end
r=k*c(1)*c(2)*(1-1/Ke*(c(3)*c(4))/(c(1)*c(2)));
for j=1:Nc
   dc(j)=Q/Vr*(cxin(j)-c(j))+ni(j)*r;
end
dc=dc';
end

function [fc] = nle_ex4_06(c)
global Vr Q k Ke cin Nc ni
r=k*c(1)*c(2)*(1-1/Ke*(c(3)*c(4))/(c(1)*c(2)));
for j=1:Nc
   fc(j)=Q/Vr*(cin(j)-c(j))+ni(j)*r;
end
end

Part 1
cs =
   0.3643 0.0643 1.1357 1.2357
Part 2
```

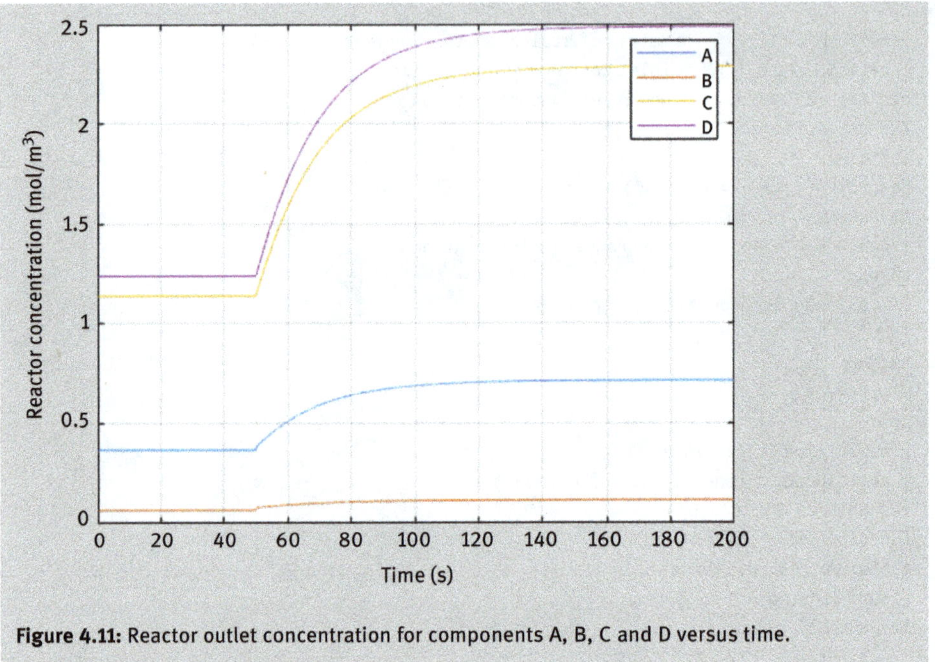

Figure 4.11: Reactor outlet concentration for components A, B, C and D versus time.

Example Single equilibrium reaction, constant volume, liquid phase, constant flow rate, isothermal CSTR reactor, stationary conditions

In an isothermal CSTR reactor, a single equilibrium reaction occurs among four components A, B, C and D as in the scheme that follows:

$$A + B \longleftrightarrow C + D$$

The kinetic of this reversible reaction is expressed by second-order relation of the type:

$$r = kC_A C_B \left[1 - \frac{1}{K_e} \frac{C_C C_D}{C_A C_B}\right]$$

The kinetic and equilibrium constants have, respectively, the values: 9 $m^3/(\text{mol s})$ and 82; the reactor volume is 6 $m^3$and the inlet concentrations of the four components are 1.5, 1.2, 0 and 0.1 mol/m^3.

Develop a Matlab code that can solve the system of nonlinear equations representing mass balances related to the four components involved in the reaction:

$$\frac{Q}{V_r}(C_{i,\text{IN}} - C_i) + \sum_{j=1}^{N_r} v_{j,i} r_j = 0 \qquad i = \text{A, B, C, D}$$

Solve the system of algebraic equations for different values of volumetric flow rate in the range $Q = 0.01$–$50\ m^3/s$ and draw two plots in which outlet reactor concentration and A conversion are reported as a function of the volumetric flow rate.

Matlab code for the solution of this example and results are as follows Figure 4.12:

```
% example 4.7
clc,clear
global Vr Q k Ke cin Nc ni

Nc=4;                  % number of components (-)
Nr=1;                  % number of reactions (-)
Vr=6;                  % reactor volume (m3)
Ke=82;                 % equilibrium constant (-)
k=9;                   % kinetic constant (m3/(mol s))
ni=[-1 -1 +1 +1];      % matrix of stoichiometric coefficients (-)
cin=[1.5 1.2 0 0.1 ]; % inlet concentrations (mol/s)

Qx=0.001:0.05:10;

%% stationary solutions
cs0=cin;
for j=1:length(Qx)
  Q=Qx(j);
  cs=fsolve('nle_ex4_07',cs0);
  Cx(j,:)=cs;
  Xa(j)=(cin(1)-cs(1))/cin(1);
end

%% plot
subplot(1,2,1)
plot(Qx,Cx)
grid
xlabel('Flow rate (m3/s)')
ylabel('Reactor outlet concentrations (mol/m3)')
legend('A','B','C','D')

subplot(1,2,2)
plot(Qx,Xa)
grid
xlabel('Flow rate (m3/s)')
ylabel('Conversion of A (-)')

function [fc] = nle_ex4_07(c)
global Vr Q k Ke cin Nc ni
r=k*c(1)*c(2)*(1-1/Ke*(c(3)*c(4))/(c(1)*c(2)));
 for j=1:Nc
  fc(j)=Q/Vr*(cin(j)-c(j))+ni(j)*r;
 end
end
```

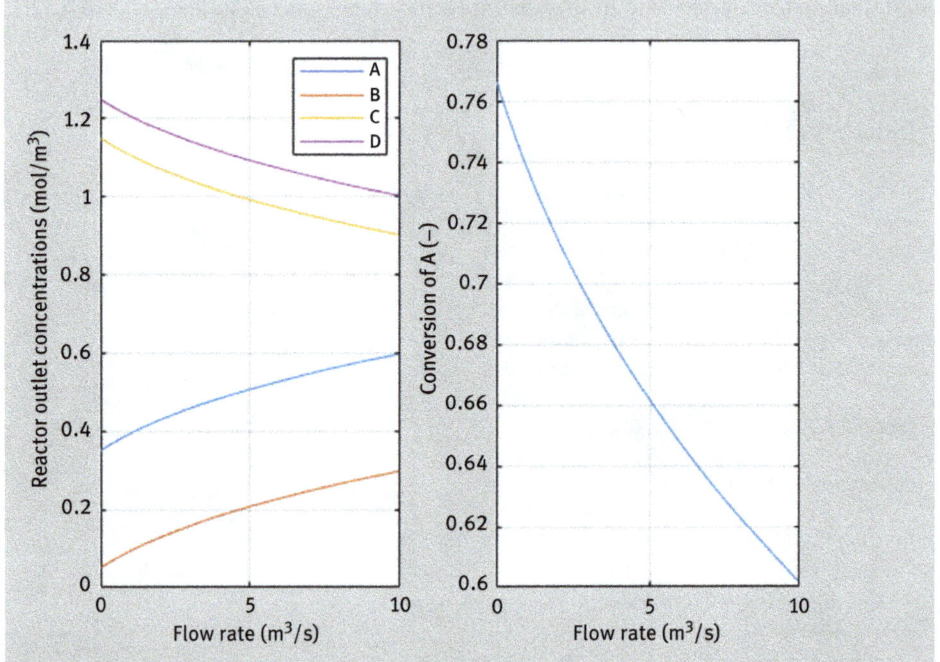

Figure 4.12: (Left) Reactor outlet concentration of components A, B, C and D versus overall volumetric feed flow rate; (right) fractional conversion of A versus overall volumetric feed flow rate.

4.4.2 Energy balance for a CSTR reactor

For what concerns energy balance related to a CSTR device illustrated in Figure 4.10 (b), the dynamic thermal behavior is represented by the following ODE:

$$m\overline{C_p}\frac{dT}{dt} = Q_{\text{IN}}\rho_{\text{IN}}C_{p,\text{IN}}T_{\text{IN}} - Q_{\text{OUT}}\rho_{\text{OUT}}C_{p,OUT}T + V_r\sum_{j=1}^{N_r}\Delta H_j r_j + V_R UA(T - T_J) \tag{4.38}$$

In the case of constant volumetric flow rate, constant density and assuming an average specific heat, eq. (4.38) can be simplified as follows:

$$\frac{dT}{dt} = \frac{Q\rho(T_{\text{IN}} - T)}{m} + \frac{V_r}{m\overline{C_p}}\sum_{j=1}^{N_r}\Delta H_j r_j + \frac{V_r UA}{m\overline{C_p}}(T - T_J) \tag{4.39}$$

And, finally, the steady-state temperature can be obtained by solving a nonlinear equation resulted by setting the accumulation term to 0 in eq. (4.39) (or alternatively in (4.38)). The relation obtained is

$$\frac{Q\rho(T_{\mathrm{IN}}-T)}{m}+\frac{V_r}{m\overline{C_p}}\sum_{j=1}^{N_r}\Delta H_j r_j+\frac{V_r UA}{m\overline{C_p}}(T-T_J)=0 \tag{4.40}$$

The complete dynamical description of CSTR reactor is obtained by solving the mass–balance ODE set represented by (4.36) coupled with energy balance (4.39). On the other hand, if the system is in stationary condition, the algebraic equations system is represented by eqs. (4.37), for what concerns mass balance, coupled with energy balance (4.40).

4.4.3 Two-phase CSTR reactors

As we have seen for batch and continuous tubular reactors, also CSTR can be operated with two interacting phases, for example, gas–liquid or liquid–liquid. It is important for this reactor configuration, the identification of the reactive phase. In the case of gas–liquid CSTR, usually, the reaction occurs in the liquid phase and the contact with gas involves only a partitioning of some components. On the contrary, in the case of liquid–liquid CSTR, both the liquid phases can be involved in the reaction network. A scheme of such type of reactor is reported in Figure 4.13.

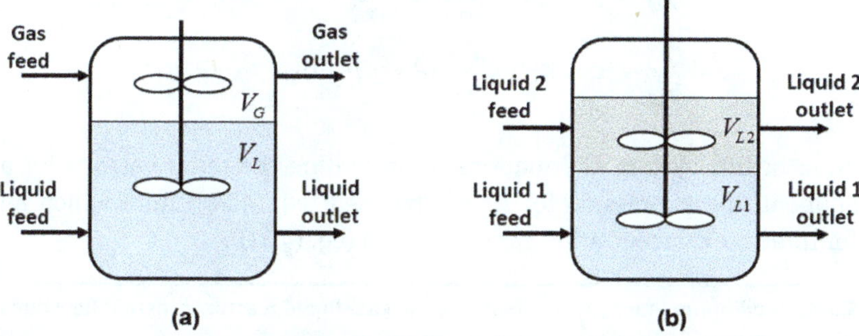

Figure 4.13: Scheme of two-phase CSTR reactors: (a) gas–liquid reactor and (b) liquid–liquid reactor.

For simplicity, reactors a and b of Figure 4.13 are both schematized without the presence of a heating jacket that, in principle, can be present on both the devices.

For a gas–liquid reactor, by assuming that the reaction occurs in the liquid phase and that reactants are transferred from gas to liquid phase, the liquid phase material balance for a generic component i is the following:

$$\frac{Q_L}{V_L}(C_{i,\mathrm{IN}}^L-C_i^L)+\sum_{j=1}^{N_r}\nu_{j,i}r_j+J_i=0 \tag{4.41}$$

While for the gas phase:

$$\frac{Q_G}{V_G}(C_{i,\mathrm{IN}}^G - C_i^G) - J_i = 0 \tag{4.42}$$

Together with relations (4.41)–(4.42), the following two relations are necessary in order to complete the gas–liquid reactor model:

$$J_i = \beta_i \left(\frac{P_i}{H_i} - C_i^L \right) \tag{4.43}$$

$$C_i^G = \frac{P_i}{R_g T} \tag{4.44}$$

Equation (4.43) represents the mass transfer flow of component i from gas to liquid, eq. while (4.44) is the definition of gas-phase concentration for the same component.

If we consider, instead, the liquid–liquid CSTR reactor schematized in Figure 4.13(b), mass balances similar to eq. (4.41) can be developed. The difference, in this configuration, consists in the possibility that different reactions can occur in both liquid phases; for the description of this system eqs. (4.46)–(4.31) are still valid and the mass balances for a generic component in both liquid phases are

$$\begin{aligned} &\frac{Q_{L1}}{V_{L1}}(C_{i,\mathrm{IN}}^{L1} - C_i^{L1}) + \sum_{j=1}^{N_{r1}} \nu_{i,j}^1 r_j \pm J_i^1 = 0 \\ &\frac{Q_{L2}}{V_{L2}}(C_{i,\mathrm{IN}}^{L2} - C_i^{L2}) + \sum_{j=1}^{N_{r2}} \nu_{i,j}^2 r_j \pm J_i^2 = 0 \end{aligned} \tag{4.45}$$

For the solution of this system of equations, representing the mass balance for a generic component, the expression for J must be used (eq. (4.26)) from which an expression for interface concentration can be derived (eq. (4.31)).

Example 4.8 Single equilibrium reaction, constant volume, gas–liquid reactor, constant flow rates (gas and liquid), isothermal CSTR reactor, stationary conditions

In an isothermal gas–liquid CSTR reactor, a single equilibrium reaction occurs in the liquid phase among four components A, B, C and D as in the scheme that follows:

$$\mathrm{A} + \mathrm{B} \longleftrightarrow \mathrm{C} + \mathrm{D}$$

The kinetic of this reversible reaction is expressed by second-order relation of the type:

$$r = kC_\mathrm{A}C_\mathrm{B}\left[1 - \frac{1}{K_e}\frac{C_\mathrm{C}C_\mathrm{D}}{C_\mathrm{A}C_\mathrm{B}}\right]$$

Two feed streams are fed to the reactor that are, respectively, a gas and a liquid stream. The only component that is partitioned between the two phases is component B that is initially present mainly in gaseous stream.

The kinetic and equilibrium constants have, respectively, the values: 90 m^3/(mol s) and 820; the reactor volume (liquid) is 60 m^3 and the gas volume is 100 m^3. The inlet concentrations of the four components in liquid stream are: 1.5, 0.1, 0 and 0.1 mol/m^3. In the entering gas stream, only component B is present, and its concentration can be calculated from temperature (450 K) and pressure (0.3 atm). The volumetric flow rates for liquid and gas feed streams are 0.2 and 3 m^3/s. The partition constant for component B, H_B, is 0.3 (atm m^3/mol).

Develop a Matlab code that can solve the system of nonlinear equations representing mass balances related to the four components involved in the liquid phase reaction:

$$\frac{Q_L}{V_L}(C_{i,\mathrm{IN}}^L - C_i^L) + \sum_{j=1}^{N_r} \nu_{j,i} r_j + J_i = 0 \quad i = \mathrm{A, B, C, D}$$

While for the gas phase:

$$\frac{Q_G}{V_G}(C_{i,\mathrm{IN}}^G - C_i^G) - J_i = 0 \qquad i = B$$

Part 1

Solve the system of five NLAEs by assigning to the mass transfer coefficient β (related to component B) the value of 0.01 s^{-1}.

Part 2

Repeat the calculation described in part 1 by varying the value of mass transfer coefficient from for 0.001 to 0.2 and draw a plot in which the fractional conversion of A in reported as a function of β.

Matlab code for the solution of this example and results are as follows:

```
% example 4.8
clc,clear
global VL QL VG QG k Ke cinL cinG Nc ni Nr Pb Hb T betaB

Nc  = 4;                     % number of components (-)
Nr  = 1;                     % number of reactions (-)
VL  = 60;                    % reactor liquid volume (m3)
VG  = 100;                   % reactor gas volume (m3)
Ke  = 820;                   % equilibrium constant (-)
k   = 90;                    % kinetic constant (m3/(mol s))
QL  = 0.2;                   % liquid feed flow rate (m3/s)
QG  = 3;                     % gas feed flow rate (m3/s)
Pb  = 0.3;                   % B partial pressure in gas (atm)
Hb  = 3;                     % GL equilibrium constant for B (atm m3/mol)
T   = 450;                   % temperature (K)
betaB= 0.01;                 % GL mass transfer coeff. for B (1/s)

Rgas = 0.08205e              % gas constant (atm m3/(mol K))
CbG = Pb/(Rgas*T);
ni =[-1 -1 +1 +1 ];          % matrix of stoichiometric coefficients (-)
cinL =[1.5 0.01 0 0.1 ]; % inlet concentrations in liquid (mol/m3)
cinG =[ 0 CbG 0 0 ];         % inlet gas concentration (mol/m3)
```

```
%% stationary solutions
c0=[cinL cinG(2)];
[cs, fs]=fsolve('nle_ex4_08',c0);

disp(' inlet outlet')
disp('------------------------------------------------------------')
fprintf(' Liquid phase : conc. A (mol/m3) %8.4f % 8.4f \n',cinL(1),cs(1))
fprintf(' Liquid phase : conc. B (mol/m3) %8.4f % 8.4f \n',cinL(2),cs(2))
fprintf(' Liquid phase : conc. C (mol/m3) %8.4f % 8.4f \n',cinL(3),cs(3))
fprintf(' Liquid phase : conc. D (mol/m3) %8.4f % 8.4f \n',cinL(3),cs(4))
disp(' ')
fprintf(' Gas phase : conc. B (mol/m3) %8.4f % 8.4f \n',cinG(2),cs(5))

%% influence of beta on A conversion
bx=0.001:0.001:0.2;
for j=1:length(bx)
  betaB=bx(j);
  c0=[cinL cinG(2)];
  [cs, fs]=fsolve('nle_ex4_08',c0);
  xA(j)=(cinL(1)-cs(1))/cinL(1);
end

plot(bx,xA)
grid
xlabel('beta comp. B')
ylabel('Conversion of A')

function [fc] = nle_ex4_08(c)
global VL QL VG QG k Ke cinL cinG Nc ni Nr Pb Hb T betaB

cL=c(1:4);
cG=c(5);
Jb=betaB*(Pb/Hb-cL(2));
r=k*cL(1)*cL(2)*(1-1/Ke*(cL(3)*cL(4))/(cL(1)*cL(2)));
Jx = [0 Jb 0 0];
 for j=1:Nc
   fc(j)=QL/VL*(cinL(j)-cL(j))+ni(j)*r + Jx(j);
 end
fc(5)=QG/VG*(cinG(2)-cG) - Jb;
end
Part 1
                    inlet    outlet
--------------------------------------------------
Liquid phase : conc. A (mol/m3)  1.5000 1.1906
Liquid phase : conc. B (mol/m3)  0.0100 0.0001
Liquid phase : conc. C (mol/m3)  0.0000 0.3094
Liquid phase : conc. D (mol/m3)  0.0000 0.4094
Gas phase :    conc. B (mol/m3)  8.1251 8.0918
```

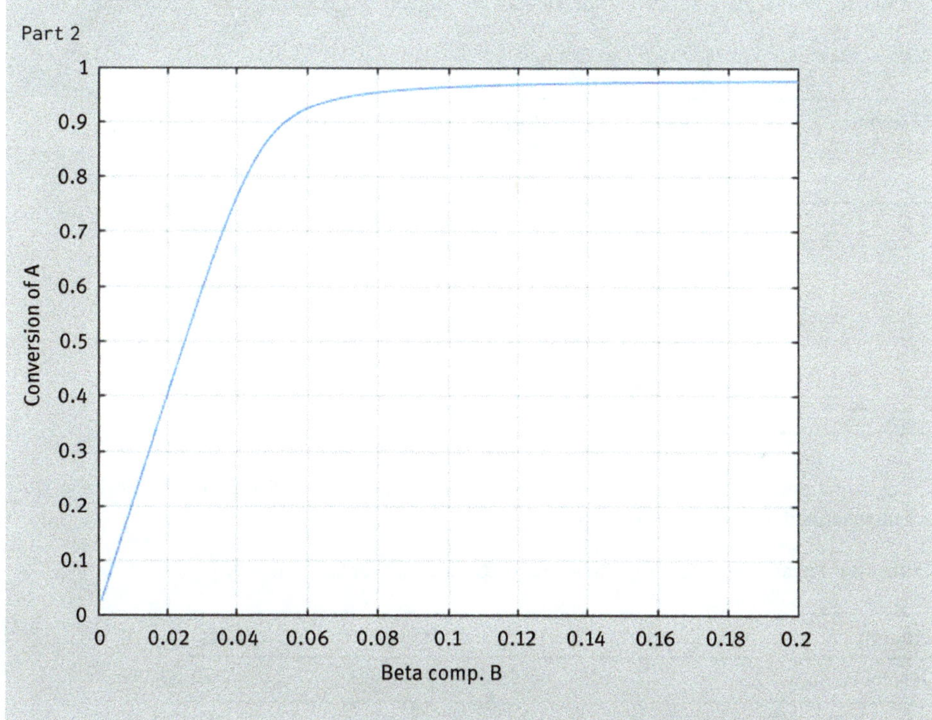

Figure 4.14: Fractional conversion of component A versus mass transfer coefficient of component B.

Example 4.9 CSTR reactor, liquid–liquid system, multiple reactions, constant feed flow rates, isothermal reactor

Two liquid phases are fed to a liquid–liquid CSTR reactor: aqueous phase and organic phase. The CSTR reactor is operated with a constant flow rate of both liquid phases and in isothermal conditions. Two chemical reactions occur in two liquid phases as follows:

$$A + B \rightarrow C \quad \text{aqueous phase}$$
$$B + D \rightarrow E \quad \text{organic phase}$$

According to this reaction scheme, component B is the only substance that is partitioned between the two contacting liquid phases. The kinetics laws of the reactions are both represented by a second-order expression as in the following table:

Table 4.12: Rate expression and kinetic parameters for Example 4.9.

Reaction	Rate expression	k
1 (aq.)	$r_1 = kC_A C_B$	8.0e−2 (m^3/(mol s))
2 (org.)	$r_2 = kC_B C_D$	2.0e−1 (m^3/(mol s))

The feed to the reactor, respectively, for aqueous and organic phases, is summarized in Table 4.13:

Table 4.13: Molar fees flow rated for Example 4.9.

Component	Aq. phase concentration (mol/m^3)	Org. phase concentration (mol/m^3)
A	1.5	0
B	1.2	0.1
C	0	0
D	0	1.8
E	0	0

Other useful data regarding reactor characteristics and physicochemical properties of the mixture are summarized in the following table.

Table 4.14: Physicochemical properties and reactor settings for Example 4.9.

Property	Value	Units
Reactor volume	3	(m^3)
Mass transfer coeff. for B, aq. phase Mass transfer coeff. for B, org. phase	0.6 0.8	(s^{-1}) (s^{-1})
Liq.–liq. partition coefficient for B	0.3	(–)
Liquid feed flow rate (aqueous phase)	0.2	(m^3/s)
Liquid feed flow rate (organic phase)	0.5	(m^3/s)

The mass balance for components in aqueous and organic phases can be written as

$$\frac{Q_{aq}}{V_{aq}}(C_{i,IN}^{aq} - C_i^{aq}) + \sum_{j=1}^{N_{raq}} v_{i,j}^{aq} r_j^{aq} \pm J_i^{aq} = 0$$

$$\frac{Q_{org}}{V_{org}}(C_{i,IN}^{org} - C_i^{org}) + \sum_{j=1}^{N_{rorg}} v_{i,j}^{org} r_j^{org} \pm J_i^{org} = 0$$

The mass transfer flow J is defined only for the components that are partitioned between the two liquid phases B in the case of our calculation.

Part 1

Solve the material balances reported above for the five components involved in the overall reactive system and calculate the composition of the two liquid exit streams. After the solution of the nonlinear equation system, check the mass balance related to the B component.

Part 2
Produce a plot in which the conversion of A is reported as a function of the mass transfer coefficients (assumed equals for aqueous and organic phases) in the range 0.001 5 s^{-1}.

Matlab code for the solution of this example and results are as follows:

```
% example 4.9
clc,clear
global VR QA QO k1 k2 cinA cinO HB betaa betao

VR    = 3;              % overall reactor volume (m3)
k1    = 8e-2;           % kinetic constant reaction 1 (m3/(mol s))
k2    = 2e-1;           % kinetic constant reaction 2 (m3/(mol s))
QA    = 0.2;            % feed flow rate of aqueous phase (m3/s)
QO    = 0.5;            % feed flow rate of organic phase (m3/s)
betaa = 0.6;            % mass transfer coeff. - aq. side (1/min)
betao = 0.8;            % mass transfer coeff. - org. side (1/min)
HB = 0.3;               % partition coefficient comp. B A/O (-)
cinA =[1.5 1.2 0 0 0 ]; % inlet conc. aq. phase (mol/m3)
cinO =[0 0.1 0 1.8 0 ]; % inlet conc. org. phase (mol/m3)

%% stationary solutions
c0=[cinA cinO];
[cs, fs]=fsolve('nle_ex4_09',c0);
fs
disp('                            inlet     outlet')
disp('-----------------------------------------------------------')
fprintf(' Aqueous phase : conc. A (mol/m3) %8.4f % 8.4f \n',cinA(1),cs(1))
fprintf(' Aqueous phase : conc. B (mol/m3) %8.4f % 8.4f \n',cinA(2),cs(2))
fprintf(' Aqueous phase : conc. C (mol/m3) %8.4f % 8.4f \n',cinA(3),cs(3))
fprintf(' Aqueous phase : conc. D (mol/m3) %8.4f % 8.4f \n',cinA(3),cs(4))
fprintf(' Aqueous phase : conc. E (mol/m3) %8.4f % 8.4f \n',cinA(3),cs(5))
disp(' ')
fprintf(' Organic phase : conc. A (mol/m3) %8.4f % 8.4f \n',cinO(1),cs(6))
fprintf(' Organic phase : conc. B (mol/m3) %8.4f % 8.4f \n',cinO(2),cs(7))
fprintf(' Organic phase : conc. C (mol/m3) %8.4f % 8.4f \n',cinO(3),cs(8))
fprintf(' Organic phase : conc. D (mol/m3) %8.4f % 8.4f \n',cinO(4),cs(9))
fprintf(' Organic phase : conc. E (mol/m3) %8.4f % 8.4f \n',cinO(5),cs(10))
disp(' ')

Bin=QA*cinA(2)+QO*cinO(2);
Bout=QA*cs(2)+QO*cs(7)+QA*cs(3)+QO*cs(10);
fprintf(' Closure on B balance : %10.6e \n',Bin-Bout)
%% influence of beta on A conversion
bx=0.001:0.001:5;
for j=1:length(bx)
  betaa=bx(j);
```

```
  betao=bx(j);
  c0=[cinA cinO];
  [cs, fs]=fsolve('nle_ex4_09',c0);
  xA(j)=(cinA(1)-cs(1))/cinA(1);
end

plot(bx,xA)
grid
xlabel('beta comp. B')
ylabel('Conversion of A')

function [fc] = nle_ex4_09(c)
global VR QA QO k1 k2 cinA cinO HB betaa betao

Va=VR*QA/(QA+QO);
Vo=VR*QO/(QA+QO);
Ca=c(1:5);
Co=c(6:10);

CAa=Ca(1);
CBa=Ca(2);
CCa=Ca(3);
CDa=Ca(4);
CEa=Ca(5);

CAo=Co(1);
CBo=Co(2);
CCo=Co(3);
CDo=Co(4);
CEo=Co(5);

r1=k1*CAa*CBa;
r2=k2*CDo*CBo;

CBoS=(betaa*Va*CBa + betao*Vo*CBo) / . . .
   (betao*Vo + betaa*Va*HB);
CBaS=HB*CBoS;
JBa=betaa*(CBa-CBaS);
JBo=betao*(CBoS-CBo);

%% material balance aqueous phase
fa(1)=QA/Va*(cinA(1)-CAa)-r1;
fa(2)=QA/Va*(cinA(2)-CBa)-r1-JBa;
fa(3)=QA/Va*(cinA(3)-CCa)+r1;
fa(4)=QA/Va*(cinA(4)-CDa);
fa(5)=QA/Va*(cinA(5)-CEa);

%% material balance organic phase
fo(1)=QA/Va*(cinO(1)-CAo);
```

```
fo(2)=QA/Va*(cinO(2)-CBo)-r2+JBo;
fo(3)=QA/Va*(cinO(3)-CCo);
fo(4)=QA/Va*(cinO(4)-CDo)-r2;
fo(5)=QA/Va*(cinO(5)-CEo)+r2;

%% merge balances
fc=[fa';fo'];

end
                        inlet    outlet
-------------------------------------------
Aqueous phase : conc. A (mol/m3)  1.5000 1.3417
Aqueous phase : conc. B (mol/m3)  1.2000 0.3440
Aqueous phase : conc. C (mol/m3)  0.0000 0.1583
Aqueous phase : conc. D (mol/m3)  0.0000 0.0000
Aqueous phase : conc. E (mol/m3)  0.0000 0.0000
Organic phase : conc. A (mol/m3)  0.0000 0.0000
Organic phase : conc. B (mol/m3)  0.1000 0.1609
Organic phase : conc. C (mol/m3)  0.0000 0.0000
Organic phase : conc. D (mol/m3)  1.8000 1.5818
Organic phase : conc. E (mol/m3)  0.0000 0.2182
Closure on B balance : -4.440892e-15
>>
```

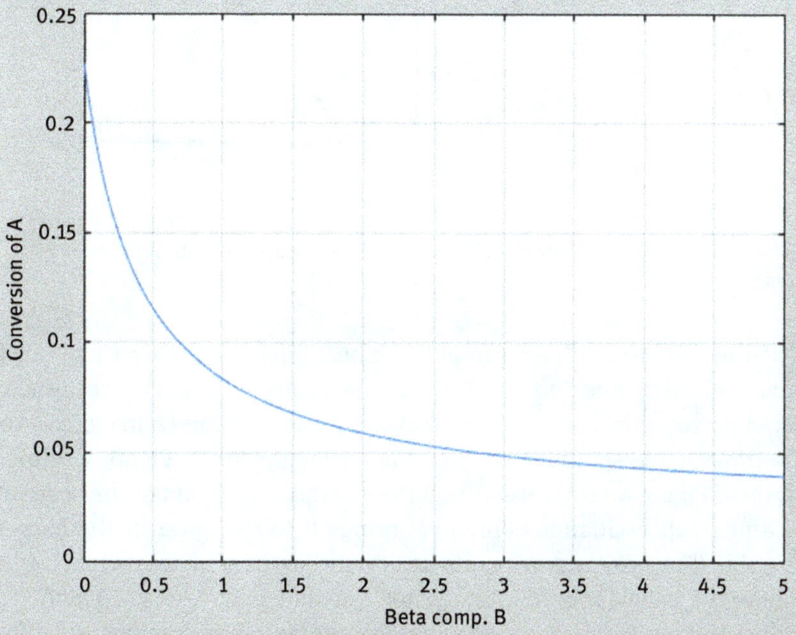

Figure 4.15: Fractional conversion of A versus mass transfer coefficient of component B.

4.5 Numerical oscillations and instabilities in a CSTR reactor

The CSTR reactor, in certain conditions, could exhibit a peculiar behavior involving more than one stationary state also in the very simple case of a single first order reaction. This aspect can be better emphasized if we consider the coupled mass and energy balance equations in the following simplified form related to a single first-order reaction A→B:

$$\begin{aligned} &QC_{A,\mathrm{IN}} - QC_{\mathrm{A}} = VkC_{\mathrm{A}} \\ &Q\rho C_p(T - T_0) - UA(T - T_J) = -VkC_{\mathrm{A}}(\Delta H) \\ &k = k_0 \exp(-Ea/RT) \end{aligned} \tag{4.46}$$

The two terms on the left-hand side of the thermal balance represent the heat removed by the flow and by the cooling jacket. The term on the right side represents, on the contrary, the heat generated by the reaction. These terms can be plotted against the temperature of the system, T, and the trend is qualitatively reported in Figure 4.16.

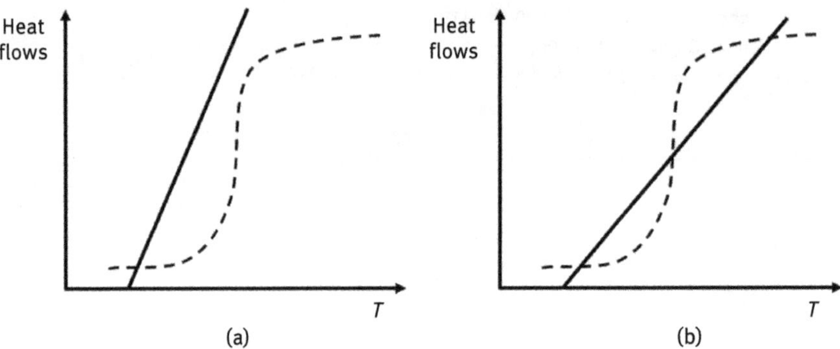

Figure 4.16: Scheme of thermal behavior of CSTR reactor: (a) single steady-state solution and (b) multiple steady-state solutions.

In Figure 4.16, the continuous line represents the heat removed while dashed line is related to heat evolved by reaction. The intersection points of the curves correspond to the steady-state solutions. In Figure 4.16(a), only one solution characterizes the system while three stable solutions are possible for the system in Figure 4.16(b). A deeper analysis of Figure 4.16(b) reveals that the intermediate temperature is an unstable point and small oscillations around it brings it to the lower or the highest temperature point. The two extreme points are, on the contrary, stable solutions.

In principle, the behavior of a CSTR reactor can shift from case a to b and vice versa according to the system parameters that are adopted. For example, the effect of adopting different values for inlet temperature T^{o} on system multiplicity is qualitatively reported in Figure 4.17.

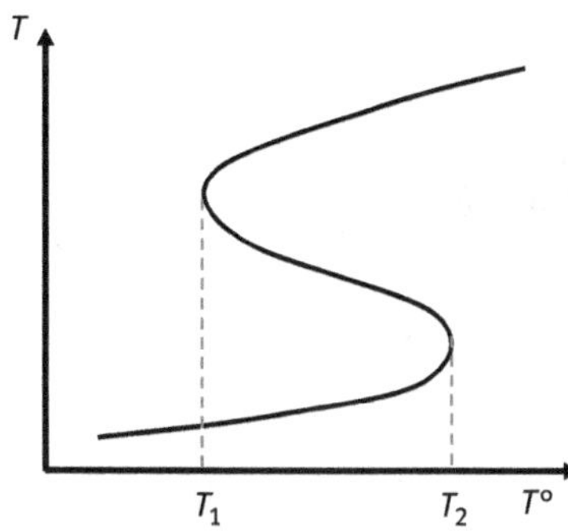

Figure 4.17: Effect of different values of inlet temperature $T°$ on system multiplicity.

By observing the trend reported in Figure 4.17, the following considerations can be established. If we adopt an inlet temperature below temperature T_1 and above temperature T_2, a single stable solution could be expected for the system. On the contrary, a value of the inlet temperature comprised in these two temperatures' interval will result in an unstable system characterized by three steady-state solutions.

A more detailed analysis on reactor oscillations and instabilities, particularly for dynamic reactor models, is outside the scope of this book and the reader is encouraged to consult dedicated books and papers.

Example 4.10 CSTR reactor, liquid phase, single reaction, constant feed flow rate, nonisothermal reactor, multiple steady states

A CSTR reactor equipped with a cooling jacket is used to perform a single liquid-phase reaction with a simple stoichiometry: A → B. The reactor is fed with a constant flow rate and the parameters for the system are summarized in Table 4.15.

Table 4.15: Physicochemical properties and reactor settings for Example 4.10.

Property	Value	Units
Volumetric feed flow rate	16.667	(cm^3/s)
Reactor volume	400	(cm^3)
Reaction mixture density	1	(g/cm^3)
Reactor inlet temperature	60	(°C)
Reactor inlet concentration	0.005	(mol/cm^3)
Preexponential factor	8.0e11	(s^{-1})
Activation energy	100,320	(J/mol)
Heat of reaction	−125,400	(J/mol)

Table 4.15 (continued)

Property	Value	Units
Specific heat of reacting mixture	4.187	(J/(g °C))
Heat exchange area	30	(cm^2)
Global heat exchange coefficient	0.3	(J/(cm^2 °C s))
Temperature of fluid in cooling jacket	50	(°C)

The mass and energy balances for component A can be written as

$$QC_{A,\mathrm{IN}} - QC_A = VkC_A$$

$$Q\rho C_p(T - T_0) - UA(T - T_j) = -VkC_A(\Delta H)$$

$$k = k_0 \exp(-Ea/RT)$$

Part 1
Solve the material and energy balances reported above using the parameters reported in the table. Print the three values for the temperature of the reactor.

Part 2
Draw a first subplot in which the left- and the right-hand side terms in the heat balance are reported as function reactor temperature T.

Part 3
Vary the inlet temperature T_0 from 0 to 100 °C and in correspondence of each value of inlet temperature draw in a second subplot the root(s) (one or more) of the energy balance.

Matlab code for the solution of this example and results are as follows:

```
% example 4.10
clc
clear
global Q ro cp T0 U A Tj V Ea k0 C0 DH

Q  = 1000/60;   % volumetric flow rate (cm3/s)
ro = 1;         % reaction mixture density (g/cm3)
V  = 400;       % reactor volume (cm3)
T0 = 60;        % reactor inlet temperature (°C)
C0 = 0.005;     % reactor inlet conc. of A (mol/cm3)
k0 = 8.0e11;    % pre-exponential factor (1/s)
Ea = 100320;    % activation energy (J/mol)
DH = -125400;   % heat of reaction (J/mol)
```

```
cp  = 4.187;  % specific heat of reaction mix. (J/(g °C))

A  = 30;      % heat exchange area (cm2)
U  = 0.3;     % global heat exchange coeff. (J/(cm2 s °C))
Tj  = 50;     % cooling jacket temperature (°C)

%% search for all the root of energy balance
RRT=rootbal(T0)

%% plot the two terms of energy balance
subplot(1,2,1)
Tx=0:400;
for j=1:length(Tx)
   [Qe(j), Qr(j)]=term(Tx(j));
end
plot(Tx,Qe,Tx,Qr)
grid
xlabel('T (°C)')
ylabel('Heat flow (J/s)')

%% plot funzione T0
subplot(1,2,2)
T0x=0:100;
for j=1:length(T0x)
  RR=rootbal(T0x(j));
  Tm=ones(length(RR),1)*T0x(j);
  plot(Tm,RR,'ro')
  xlabel('Inlet temperature (°C)')
  ylabel('Reactor temperature (°C)')
  grid
  hold on
  drawnow
end

function [fT] = nle_ex4_10(T)
global Q ro cp T0 U A Tj V Ea k0 C0 DH

Q1 = Q*ro*cp*(T-T0);
Q2 = U*A*(T-Tj);
Tk = T+273;
R = 1.987*4.187;
k = k0*exp(-Ea/(R*Tk));
num = -Q*C0*V*(DH)*k;
den = Q+V*k;
fT=Q1+Q2-num/den;
end
```

```
function [rootT] = rootbal(T0x)
%-------------------------------------
global T0
T0=T0x;
Tx=0:400;
nR=0;
for j=1:length(Tx)-1
  TA=Tx(j);
  TB=Tx(j+1);
  fTA=nle_ex4_10(TA);
  fTB=nle_ex4_10(TB);
  chks=fTA*fTB;
  if chks<0
    Tr=fzero('nle_ex4_10',[TA TB]);
    nR=nR+1;
rootT(nR)=Tr;
end

end
end

function [Qe, Qr] = term(T)
%--------------------------------------
global Q ro cp T0 U A Tj V Ea k0 C0 DH
Q1 = Q*ro*cp*(T-T0);
Q2 = U*A*(T-Tj);
Tk = T+273;
R = 1.987*4.187;
k = k0*exp(-Ea/(R*Tk));
num = -Q*C0*V*(DH)*k;
den = Q+V*k;

Qe=Q1+Q2;
Qr=num/den;
end

RRT =
59.2993 118.7549 190.1105
```

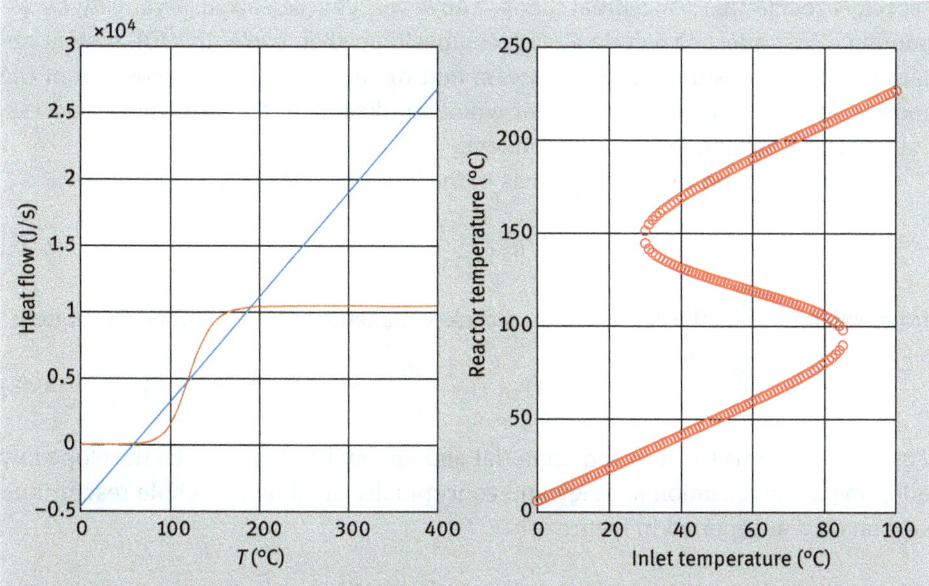

Figure 4.18: (Left) Heat flow rate versus reactor temperature; (right) reactor temperature versus feed temperature.

4.6 PFR reactor with recycle

In some specific situations, the reactor (PFR, CSTR or others) can be operated inside a circulation loop by splitting the outlet stream into two portions and sending back one of the resulting streams joining the fresh feed to the reactor. A general scheme of this type of configuration, referred to a PFR reactor, is reported in Figure 4.19 [8].

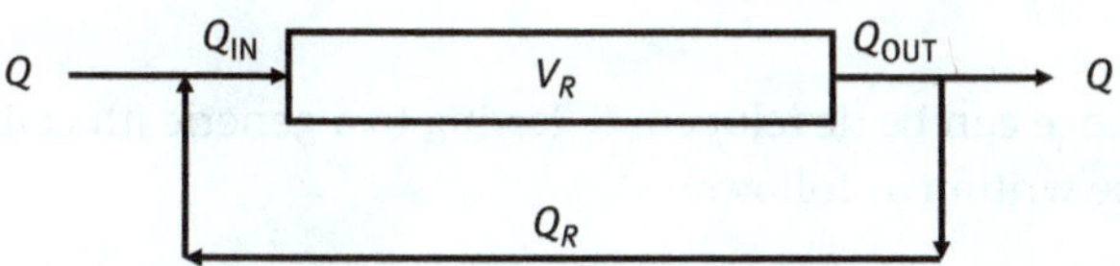

Figure 4.19: Scheme of a PFR reactor with a recirculating stream.

The stationary mass and energy balances related to recycled PFR reactors are formally the same as developed for the reactor without recycle stream and are represented by eqs. (4.6) and (4.22). The peculiarity consists in the fact that the ODEs (coupled mass and energy balances) cannot be integrated along reactor volume as the inlet conditions are not known because they depend, in part, on the outlet recycle stream. This conceptual "loop," however, can be solved iteratively by assuming a first guest of recycle stream composition; then solve the ODE system obtaining the composition of outlet stream and finally adjust the composition of the recycle stream. A solver of nonlinear equation ("fzero" in Matlab) can be successfully used for solving this type of problem.

The recycle ratio can be defined as in the following relation:

$$R_S = \frac{Q}{Q + Q_R} \tag{4.47}$$

from which if we fix the recycle ratio, recycle volumetric flow rate can be calculated:

$$Q_R = Q\frac{(1 - R_S)}{R_S} \tag{4.48}$$

For what concerns the transient material and energy balances can be developed by adopting a discretization strategy and conceptually dividing the whole reactor into several cells as reported in Figure 4.20.

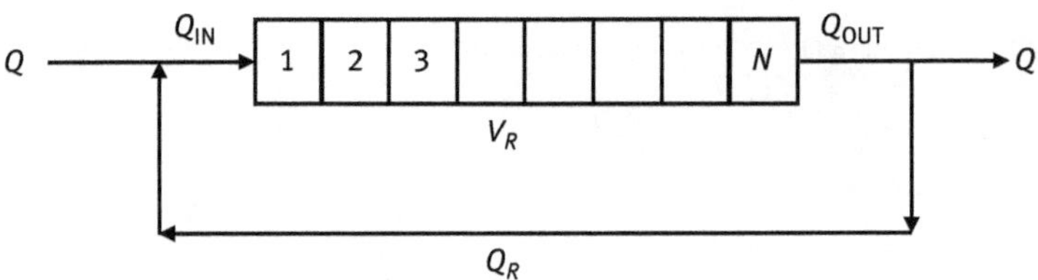

Figure 4.20: Scheme of a recirculated PFR reactor discretized into cells.

Assuming that the behavior of each cell can be described as a CSTR reactor, the mass balance for a generic jth cell is similar to eq. (4.36):

$$\frac{dC_i^j}{dt} = \frac{Q}{V_j}(C_i^{j-1} - C_i^j) + \sum_{k=1}^{N_r} \nu_{k,i} r_k^j \tag{4.49}$$

In a similar way the energy balance can be developed. Referring to a generic jth cell a relation similar to (4.39) can be written as follows:

$$\frac{dT_j}{dt} = \frac{Q\rho(T_{j-1} - T_j)}{m_j} + \frac{V_j}{m_j\overline{C_p}}\sum_{k=1}^{N_r} \Delta H_{r,k} r_k^j + \frac{V_j U_j A_j}{m_j \overline{C_p}}(T_j - T_{\text{Jaket}}) \tag{4.50}$$

For a tubular reactor that has been divided into N separate CSTR-like cells, is Nc components are involved in the operation, the overall number of coupled initial values of ODEs (4.49) and (4.50) is $N(Nc + 1)$. This ODE system can be solved by adopting suitable initial conditions for the initial concentrations in the cells and integration at long time furnishes the steady-state solution. Moreover, in this approach, the transient behavior of each cell, and also the exit one, can be obtained. This information could result and be particularly useful in the case of reactor startup and shutdown.

Example 4.11 PFR reactor, liquid phase, single reaction, constant feed flow rate, isothermal reactor, recirculating stream, steady state

A PFR reactor is operated in isothermal conditions and is used to perform a single liquid-phase reaction with a simple stoichiometry: A → B. The reactor is fed with a constant volumetric flow-rate and the exit stream is divided in two portions: the first represents the product while the second is recycled back and joined with the fresh feed and fed to the reactor. The parameters of the system are summarized in Table 4.16.

Table 4.16: Physicochemical properties and reactor settings for Example 4.11.

Property	Value	Units
Volumetric feed flow rate	0.05	(m^3/s)
Reactor volume	3	(m^3)
Kinetic constant	0.03	(s^{-1})
Splitting ratio $R_S = Q/(Q + Q_r)$	0.1	(–)
Molar feed flow rate of A	2	(mol/s)
Molar feed flow rate of B	0	(mol/s)

The mass balances for, respectively, components A and B can be written as

$$\frac{dF_A}{dV} = -r$$

$$\frac{dF_B}{dV} = +r$$

Part 1

Solve the material and energy balances reported above using the constants reported in the table. Print the value of fractional conversion of reactant A.

Part 2

Draw a plot in which fractional conversion is reported as a function of the recycling ratio between 0 and 1.

Matlab code for the solution of this example and results are as follows:

```
% example 4.11
clc,clear
global Vr k1 Rs FA0 FB0
global Q Q1 Q2 Qr xA
Vr = 3;      % overall reactor volume (m3)
k1 = 0.03;  % kinetic constant of reaction (1/s)
Q  = 0.05;  % feed flow rate (m3/s)
Rs = 0.1;    % splitting ratio (-)
FA0 = 2;     % A feed flow rate (mol/s)
FB0 = 0;     % B feed flow rate (mol/s)
fzero('tearing',.5)
xA
Rx=0.01:0.01:0.99;
for j=1:length(Rx)
  Rs=Rx(j);
  x=fzero('tearing',0.5);
  xAA(j)=xA;
end

plot(Rx,xAA)
grid
xlabel('Recycle ratio (-)')
ylabel('A conversion (-)')

function [df, CBx]=ode_ex4_12(v,f)
global Vr k1 Rs FA0 FB0
global Q Q1 Q2 Qr
Ca=f(1)/Q1;
r=k1*Ca;
%% material balance aqueous phase
 df(1) = -r;
 df(2) = +r;
%% transposition of derivative arrays
df=df';
end

function [dFx] = tearing(FAx)
global Vr k1 Rs FA0 FB0
global Q Q1 Q2 Qr xA
Q2 = Q/Rs;
Qr = Q2-Q;
Q1 = Q2;
FAr = FAx;
FA1 = FA0 + FAr;
FA2 = FAr*Q2/Qr;
FB2 = FA1 - FA2;
```

```
FBr = FB2*Q2/Qr;
FB1 = FB0 + FBr;
F0=[FA1 FB1];
vspan=0:Vr/500:Vr;
options=odeset('RelTol',1e-13,'AbsTol',1e-13);
[v,fx]=ode15s(@ode_ex4_11,vspan,F0);
dFx = FA2 - fx(end,1);
xA = (FA1-FA2)/FA1;
end
xA =

0.1647
```

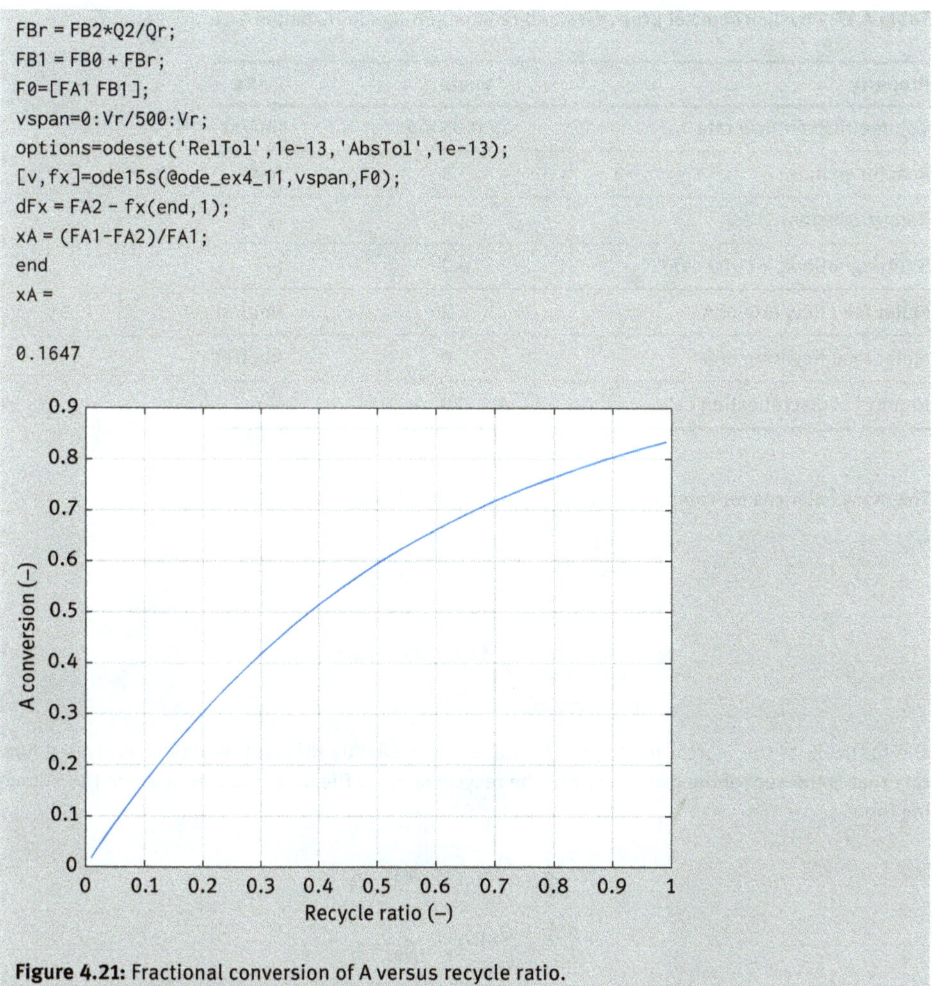

Figure 4.21: Fractional conversion of A versus recycle ratio.

Example 4.12 PFR reactor, liquid phase, single reaction, constant feed flow rate, isothermal reactor, recirculating stream, unsteady state

A PFR reactor is operated in isothermal conditions and is used to perform a single liquid-phase reaction with a simple stoichiometry: A → B. The reactor is fed with a constant volumetric flow rate and the exit stream is divided in two portions: the first represents the product while the second is recycled back and joined with the fresh feed and fed to the reactor. The purpose is to simulate the transient behavior to the reactor to reach the steady state. The parameters of the system are summarized in Table 4.17.

Table 4.17: Physicochemical properties and reactor settings for Example 4.11.

Property	Value	Units
Volumetric feed flow rate	0.05	(m^3/s)
Reactor volume	3	(m^3)
Kinetic constant	0.03	(s^{-1})
Splitting ratio $R_s = Q/(Q + Q_r)$	0.2	(–)
Molar feed flow rate of A	2	(mol/s)
Molar feed flow rate of B	0	(mol/s)
Number of discretization cells	20	(–)

The mass balances for components A and B in the cells 2–N can be written as

$$\frac{dC_A^j}{dt} = \frac{Q_{tot}}{V_{cell}}(C_A^{j-1} - C_A^j) - r_j \quad j = 2 \ldots N$$

$$\frac{dC_B^j}{dt} = \frac{Q_{tot}}{V_{cell}}(C_B^{j-1} - C_B^j) + r_j \quad j = 2 \ldots N$$

$$Q_{tot} = Q + Q_R$$

The first cell, at the reactor inlet, must be treated in a slightly different way as it receives a flow rate that is the sum of the fresh feed and the recycle stream. The balances then assume the following form:

$$\frac{dC_A^1}{dt} = \frac{Q_{tot}}{V_{cell}}(C_A^{IN} - C_A^1) - r_1$$

$$\frac{dC_B^1}{dt} = \frac{Q_{tot}}{V_{cell}}(C_B^{IN} - C_B^1) + r_1$$

Solve the material and energy balances reported above using the constants reported in the table by integrating the ODE system in the time range 0–400 s. Draw two subplots as follows: in the first one, report the steady-state solution related to the reactor that is the concentration profiles of A and B as a function of the reactor volume; in the second plot, report the profiles of the exit of the reactor (last cell) as a function of time. Verify that the reactor reaches the steady state after a certain time.

Matlab code for the solution of this example and results are as follows:

```
% example 4.12
clc,clear
global Vr k1 Rs FA0 FB0
global Q N
Vr = 3;      % overall reactor volume (m3)
k1  = 0.03; % kinetic constant of reaction (1/s)
Q = 0.05;    % feed flow rate (m3/s)
Rs = 0.2;    % splitting ratio (-)
FA0 = 2;     % A feed flow rate (mol/s)
FB0 = 0;     % B feed flow rate (mol/s)
N = 20;      % number of cells (n)
tspan=0:0.1:400;
c0=zeros(2*N,1);
[t,cx]=ode45('ode_ex4_12',tspan,c0);

%% plots
n=1:N;
Vj=Vr/N;
v=n*Vj;
cA=cx(end,1:N);
cB=cx(end,N+1:2*N);
subplot(1,2,1)
plot(v,cA,'ro-',v,cB,'bo-')
grid
xlabel('Reactor volume (m3)')
ylabel('Conc. (mol/L)')
legend('cA','cB')
subplot(1,2,2)
cAout=cx(:,N);
cBout=cx(:,2*N);
plot(t,cAout,'-r',t,cBout,'b-')
grid
xlabel('Time (s)')
ylabel('Outlet conc. (mol/L)')
legend('cA','cB')
```

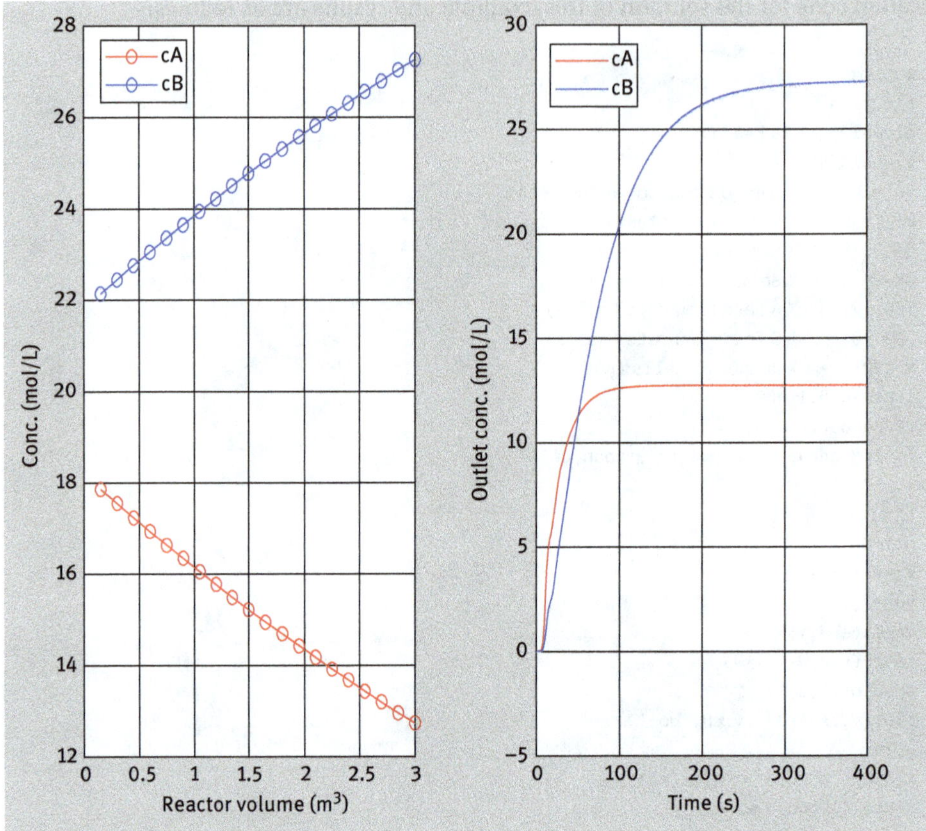

Figure 4.22: (Left) Concentration of A and B versus reactor volume in stationary conditions; (right) reactor outlet concentration of A and B versus time.

List of symbols

a_k	Liquid–liquid specific interface area	$[m^2/m^3]$
A_S	Reactor specific heat transfer area	$[m^2/m^3]$
C_i	Concentration of component *i*	$[mol/m^3]$
C_i^L	Concentration of component *i* in gas phase	$[mol/m^3]$
C_i^G	Concentration of component *i* in liquid phase	$[mol/m^3]$
C_{Ai}	Concentration of component A at inlet	$[mol/m^3]$
C_{Ae}	Concentration of component A at exit of the reactor	$[mol/m^3]$
$C_{i,\mathrm{IN}}$	Concentration of component *i* at inlet	$[mol/m^3]$
C_i^1, C_i^2	Concentration of component *i* in liquid phases 1 and 2	$[mol/m^3]$
C_i^{1*}	Equilibrium concentration of component *i* in liquid phases 1	$[mol/m^3]$
C_i^{2*}	Equilibrium concentration of component *i* in liquid phases 2	$[mol/m^3]$

C_i^*	Equilibrium concentration of component *i*	[mol/m^3]
C_{pm}	Average specific heat of reaction mixture	[J/(kg K)]
$\bar{C}_p$	Average specific heat of reaction mixture	[J/(kg K)]
$C_{p,\mathrm{IN}}$	Specific heat of the mixture at reactor inlet	[J/(kg K)]
$C_{p,\mathrm{OUT}}$	Specific heat of the mixture at the reactor outlet	[J/(kg K)]
ΔH_j	Heat of reaction related to reaction *j*	[J/mol]
Q_E	Heat exchanged from the reactor to surroundings	[J/s]
Q_{Re}	Heat of reaction term of the energy balance	[J/s]
Q_R	Volumetric flow rate of the recycling stream	[m^3/s]
$Q_{\mathrm{IN}}, Q_{\mathrm{OUT}}$	Volumetric flow rate at inlet and outlet	[m^3/s]
Q	Volumetric flow rate	[m^3/s]
U	Overall heat transfer coefficient	[J/(m^2 s K)]
A	Heat transfer area or cross-section area	[m^2]
E_A, E_{Ai}	Activation energy, activation of *i*th reaction	[J/mol]
k_i	Kinetic constant of *i*th chemical reaction	[*]
$k_{i\,\mathrm{ref}}$	Reference kinetic constant of *i*th chemical reaction	[*]
k_L^i	Liquid–liquid mass transfer coefficient of component *i*	[m/s]
m	mass	[kg]
H_i	Gas–liquid partition constant (Henry's constant)	[atm m^3/mol]
H_L^i	Liquid–liquid partition constant	[–]
R_g	Gas constant	[atm m^3/(mol K)]
V, V_R	Reactor volume	[m^3]
V_L	Liquid-phase volume	[m^3]
V_G	Gas-phase volume	[m^3]
V_1	Volume of immiscible liquid phase 1	[m^3]
V_2	Volume of immiscible liquid phase 2	[m^3]
V_k	Volume of phase *k*	[m^3]
r	Reaction rate	[mol/(m^3 s)]
P, P_o	Total pressure	[atm]
p_j, P_i	Partial pressure of component *i* or *j*	[atm]
y_i	Mole fraction of component *i*	[–]
r_k	Reaction rate of reaction *k*	[mol/(m^3 s)]
R	Reactor radius	[m]
R_S	Splitting ration for the recycled reactor	[mol/(m^3 s)]
N_C	number of Components in the mixture	[–]
N_{Ci}	number of Components involved in the reaction i	[–]
N_R	Number of independent chemical reactions	[–]
N	number of Discretization cells	[–]
T, T_o	Temperature of the reaction mixture	[K]
T_{IN}	Temperature at reactor inlet	[K]
T_J	Reactor jacket temperature	[K]
T_{ref}	Reference temperature in the Arrhenius equation	[K]
J_i	Gas–liquid mass transfer flow of component *i*	[mol/(m^3 s)]
J_1^i	Liquid–liquid mass transfer flow of component *i* (bulk 2 to interface)	[mol/(m^3 s)]
J_2^i	Liquid–liquid mass transfer flow of component *i* (interface to bulk 1)	[mol/(m^3 s)]
F_i	Molar flow rate of component *i* in feed	[mol/s]
F, F_o	Overall molar flow rate	[mol/s]

F_{Ai}	Molar flow rate of component A in feed	[mol/s]
F_{Ae}	Molar flow rate of component A at reactor outlet	[mol/s]
z	Axial coordinate in the reactor	[m]
L	Reactor length	[m]
u	Linear velocity in the reactor	[m/s]
n_i	Moles of component i	[mol]

Greek letters

ρ_j	Density of component j	[kg/m^3]
ρ_{IN}	Density of inlet mixture	[kg/m^3]
ρ_{OUT}	Density of outlet mixture	[kg/m^3]
λ_j	Reaction order for component j	[–]
β_i	Volumetric gas–liquid mass transfer coefficient for component i	[s^{-1}]
β_1^i	Volum. gas–liquid mass transfer coeff. for component i in phase 1	[s^{-1}]
β_2^i	Volum. gas–liquid mass transfer coeff. for component i in phase 2	[s^{-1}]
β_k^i	Volum. gas–liquid mass transfer coeff. for component i in phase k	[s^{-1}]
v_i	Stoichiometric coefficient of component i	[–]
$v_{i,k}$	Stoichiometric coefficient of component i in reaction k	[–]

*Units of kinetic constants depend on the expression of the reaction rate.

References

[1] H.S. Fogler. Elements of Chemical Reaction Engineering (5th edition). Prentice Hall, 2016.

[2] E.B. Nauman. Chemical Reactor Design, Optimization, and Scaleup (2th edition). Wiley, 2008.

[3] O. Levenspiel. Chemical Reaction Engineering (3th edition). Wiley: 1998.

[4] C.G. Hill. An Introduction to Chemical Engineering Kinetics and Reactors Design. John Wiley & Sons: 1977.

[5] S. Ergun. Fluid flow through packed columns. Chemical Engineering Progress 1952, 48, 89.

[6] C.J. Geankoplis. Transport Processes and Unit Operations (3th edition). Prentice-Hall, International, New Jersey.: 1993.

[7] K. Thulukkanam. Heat Exchanger Design Handbook (2th edition). CRC Press, 2013.

[8] V. Russo, T. Salmi, F. Mammitzsch, O. Jogunola, R. Lange, J. Wärnåa, J.-P. Mikkola. First, second and nth order autocatalytic kinetics in continuous and discontinuous reactors. Chemical Engineering Science 2017, 172, 453–462.

Chapter 5
Real reactors for heterogeneous catalysis

5.1 Introduction

Chemical reactors operating in continuous, as known, are characterized by nonidealities, overall concerning the flow pattern. Modeling a nonideal continuous reactor means to develop an adequate model that takes into account all the occurring chemical and physical phenomena. The following guidelines are suggested when developing models for nonideal reactors:

- *The model must be mathematically tractable*: the equations used to describe a chemical reactor should be able to be solved without an inordinate expenditure of human or computer time.
- *The model must realistically describe the characteristics of the nonideal reactor*: the phenomena occurring in the nonideal reactor must be reasonably described physically, chemically and mathematically.
- *The model must not have more than two adjustable parameters*: an expression with more than two adjustable parameters can be fitted to a great variety of experimental data, and the modeling process in this circumstance is nothing more than an exercise in curve fitting.

The nonideality of a reactor operating in flow is experimentally studied by conducting residence time distribution (RTD) experiments, determined from a tracer test, either stepwise or pulse test. In this way, it is possible to provide the existence/absence of either bypasses or dead volumes and measure the degree of backmixing in the case of a tubular reactor.

Tubular reactors may be empty or they may be packed with some material that acts as a catalyst, heat-transfer medium, promoting interphase contact. Often it is assumed that the fluid moved through the reactor in piston-like flow (PFR), and every atom spends an identical length of time in the reaction environment. Here, the velocity profile is put and there is no axial mixing. Several models can be used to describe the fluid dynamics of the system: (i) tank-in-series (TIS) model, where the parameter is the number of tanks; (ii) dispersion model: where the unknown parameter to be fitted is the dispersion coefficient D_z; (iii) laminar flow reactor (LFR). In this chapter the three models will be analyzed and treated. An effort will be made to show the readers examples useful for further implementations.

https://doi.org/10.1515/9783110632927-005

5.2 Tank-in-series model

To investigate the nonideality of a tubular reactor, one possible option is to adopt the TIS model, starting from the idea that a plug-flow reactor can be approximated to an infinite series of continuous stirred tank vessels (CSTR). To simplify the approach, a series of three CSTR is considered (see Figure 5.1), all the reactors characterized by the same volume. Consequently, the sum of the volumes of all the reactors must be equal to the volume of the reactor we are characterizing.

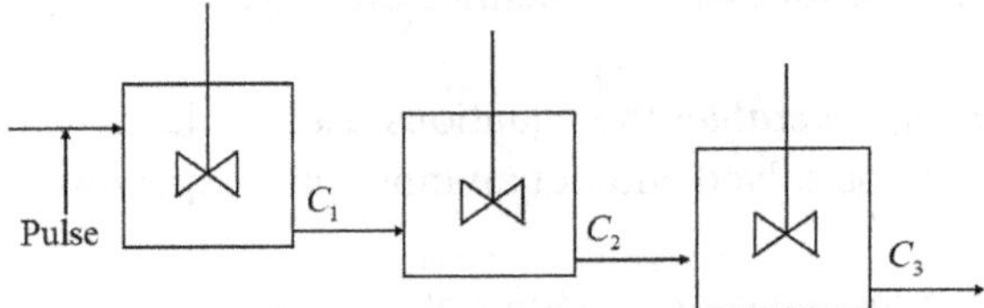

Figure 5.1: Sketch of the tank-in-series model.

By performing pulse experiments with a fixed amount of moles of tracer in the feed, n_{feed} (see details in the literature [1, 2]), it is possible to determine the experimental RTD function, $E(t)$, for the whole series:

$$E(t)\Delta t = \frac{\dot{V}}{n_{\text{feed}}} C_3(t)\Delta t = \frac{C_3(t)}{\int_0^\infty C_3(t)dt}\Delta t \rightarrow E(t) = \frac{C_3(t)}{\int_0^\infty C_3(t)dt} \tag{5.1}$$

This function can be used to obtain the concentration at the outlet of each CSTR till defining the concentration at the outlet of the reactor C_3.

For a single dynamic CSTR, it is possible to write the following mass balance:

$$V_1\frac{dC_1}{dt} = -\dot{V}C_1 \rightarrow C_1 = C_0 e^{-t/\tau_1}, \quad C_0 = \frac{n_{\text{feed}}}{V_1} = \frac{\dot{V}\int_0^\infty C_3(t)dt}{V_1} \tag{5.2}$$

If the volumetric flowrate is constant, it is possible to write the mass balance equation for the second CSTR:

$$V_2\frac{dC_2}{dt} = \dot{V}C_1 - \dot{V}C_2 \tag{5.3}$$

Substituting eq. (5.2) into eq. (5.3), it is possible to obtain the following:

$$\frac{dC_2}{dt} + \frac{C_2}{\tau_i} = \frac{C_0}{\tau_i}e^{-t/\tau_i} \rightarrow C_2(t=0) = 0 \rightarrow C_2 = \frac{C_0 t}{\tau_i}e^{-t/\tau_i} \tag{5.4}$$

Iterating the calculation for the third and final CSTR, the following equation can be obtained:

$$C_3 = \frac{C_0 t^2}{2\tau_i^2} e^{-t/\tau_i} \tag{5.5}$$

Equation (5.5) can be substituted into eq. (5.1), giving the RTD function:

$$E(t) = \frac{C_3(t)}{\int_0^\infty C_3(t)dt} = \frac{C_0 t^2 e^{-t/\tau_i}/(2\tau_i^2)}{\int_0^\infty \frac{C_0 t^2}{2\tau_i^2} e^{-t/\tau_i} dt} = \frac{t^2}{2\tau_i^3} e^{-t/\tau_i} \tag{5.6}$$

It is then easy to generalize the problem for n CSTR, obtaining

$$E(t) = \frac{t^{n-1}}{(n-1)!\tau_i^n} e^{-t/\tau_i}, \quad V = nV_i, \quad \tau = n \tag{5.7}$$

That in dimensionless form, defining a dimensionless time as $\Theta = t/\tau$, eq. (5.7) becomes

$$E(\Theta) = \tau E(t) = \frac{n(n\Theta)^{n-1}}{(n-1)!} e^{-n\Theta} \tag{5.8}$$

The implementation of eq. (5.8) in MATLAB is straightforward, as given in Example 5.1. The unique parameter to be adjusted on the experimental data is the number of CSTR, n (Figure 5.2).

Example 5.1 Tank in series
For the TIS model, implement the following equation varying Θ = 0:10–2:3 and n = 1:100:

$$E(\Theta) = \frac{n(n\Theta)^{n-1}}{(n-1)!} e^{-n\Theta}$$

Matlab code for the solution of this example and results are reported as follows Figure 5.2:

```
% example 3.1
clc, clear
t = [0:1e-2:3]' ;
n = [1:100]' ;


for k = 1:100
E(:,k) = (k.*(k.*t).^(k-1)).*exp(-k.*t)./factorial((k-1)) ;
end
```

```
figure(1)
subplot(1,2,1)
surf(n,t,E,'EdgeColor','none')
xlabel('\it \theta \rm[-]'), ylabel('\it n \rm[-]'), zlabel('\it E_{\theta} \rm[-]')
subplot(1,2,2)
plot(t,E(:,2),'-r',t,E(:,4),'-k',t,E(:,10),'-b',t,E(:,100),'-g')
xlabel('\it \theta \rm[-]'), ylabel('\it E_{\theta} \rm[-]')
legend('\it n\rm = 2','\it n\rm = 4','\it n\rm = 10','\it n\rm = 100')
grid on
```

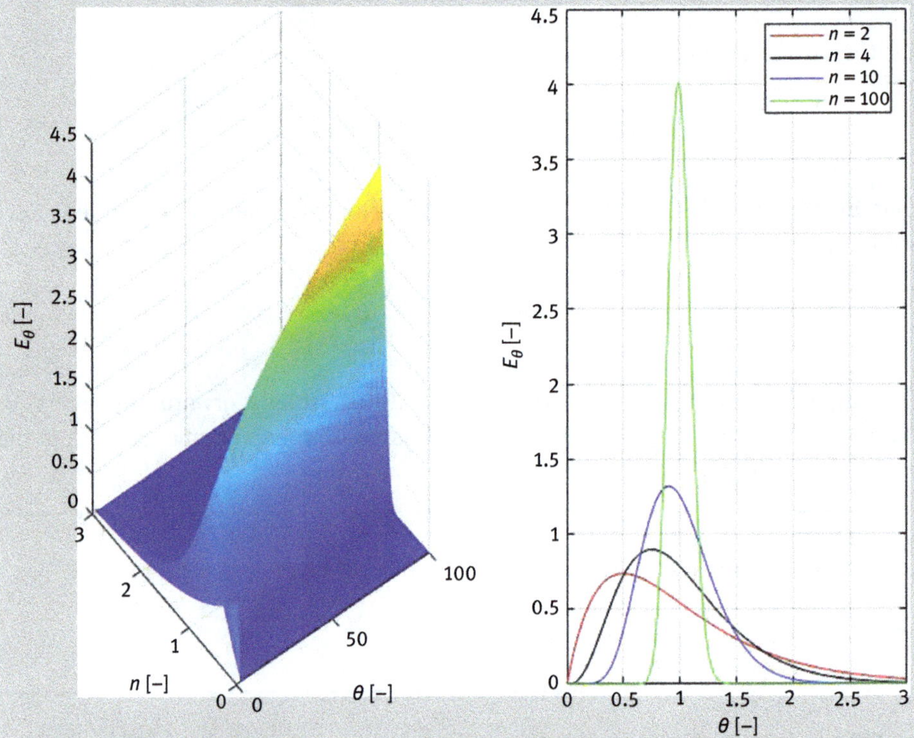

Figure 5.2: Dimensionless RTD function profile as a function of the dimensionless time and the number of CSTR (left). $E(\Theta)$ versus Θ plot for different values of the number of CSTR (right).

Higher number of CSTR leads to a sharper peak. With an infinite number, in theory the output would give a δ-Dirac function, characteristic of a plug flow.

5.3 Axial dispersion model

The axial dispersion model is based on the assumption that there is an axial dispersion of the material flowing in the continuous reactor, which is governed by an analogy to Fick's law of diffusion, superimposed on the flow. Thus, in addition to transport by bulk flow convection, every component is transported through the cross section of the reactor at a rate resulting from molecular and convective diffusion. By convective diffusion we mean either Aris–Taylor dispersion in LFR or turbulent diffusion resulting from turbulent eddies. The experimental RTD function is obtained conducting RTD experiments, wither in pulse or stepwise modality, using an inert tracer [2]. At a certain time ($t = 0$), a fixed amount of tracer is fed either in continuous (stepwise experiment) or as a pulse. In the last case, the injection time must be sufficiently short to approximate it as a δ-Dirac function (see Figure 5.3). At the outlet of the reactor ($z = L$) it is possible to obtain different results depending on the degree of the dispersion, having as higher extreme δ-Dirac function, characteristic of a plug-flow fluid dynamics, shifted in timescale by the average residence time. In the case of the occurrence of axial dispersion, the experimental output is a peak, whose broadening is proportional to the degree of backmixing (see Figure 5.3).

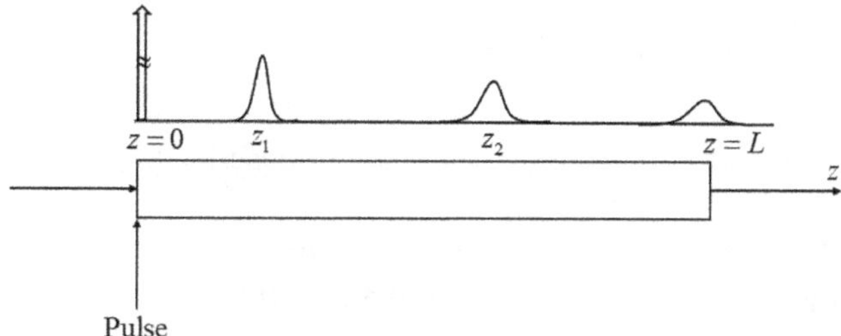

Figure 5.3: Theoretical view of a residence time distribution experiment in pulse modality for the determination of the axial dispersion coefficient.

To describe the system, the molar flowrate of the tracer is written as follows:

$$\dot{n}_\mathrm{T} = -D_z A \frac{\partial C_\mathrm{T}}{\partial z} + uAC_\mathrm{T} \tag{5.9}$$

with D_Z as the axial dispersion coefficient and u as the superficial velocity.

A mole balance on an infinitesimal length z of the reactor, written on the inter-tracer T gives the following:

$$A\frac{\partial C_\mathrm{T}}{\partial t} = -\frac{\partial \dot{n}_\mathrm{T}}{\partial z} \tag{5.10}$$

Substituting it is possible to obtain

$$\frac{\partial C_T}{\partial t} = -\frac{\partial u C_T}{\partial z} + D_z \frac{\partial^2 C_T}{\partial z^2} \tag{5.11}$$

Introducing the dimensionless quantities, $y = C_T / C_{T,\text{feed}}$, $\chi = z/L$ and $\Theta = t/\tau$, the dimensionless number called Péclet number appears and the mass balance is written as follows:

$$\frac{\partial y}{\partial \Theta} = -\frac{\partial y}{\partial \chi} + \frac{1}{\text{Pe}} \frac{\partial^2 y}{\partial \chi^2} \tag{5.12}$$

Equations (5.11) can be rewritten for a generic chemical component *i* adding a generation/consumption term to include the possibility of the occurrence of a chemical reaction:

$$\frac{\partial C_i}{\partial t} = -\frac{\partial u C_i}{\partial z} + D_z \frac{\partial^2 C_i}{\partial z^2} + \sum_k v_{ik} r_k \tag{5.13}$$

In dimensionless form, the Damkohler number (D_A) appears if a first-order kinetic rate equation is introduced:

$$\frac{\partial y}{\partial \Theta} = -\frac{\partial y}{\partial \chi} + \frac{1}{\text{Pe}} \frac{\partial^2 y}{\partial \chi^2} - D_A y \tag{5.14}$$

These last equations are not used for tracer experiments but to describe kinetic experiments conducted in nonideal tubular reactors.

In all cases, the mass balance equations have a form of partial differential equations. Thus, to solve the mathematical system, both initial and boundary conditions are needed.

When facing with RTD experiments, the initial condition is rather straightforward, as it is possible to consider the pipe empty of the tracer. For kinetic experiments, instead, everything depends on the operation modality. For instance, usually the reactor is filled with a solvent and at $t = 0$, the reactants are fed to the reactor.

In the case of boundary conditions, two different options are possible:

- *Closed–closed vessels*: plug-flow condition at the pipe entrance. At the outlet we have continuity:

$$\begin{aligned} C_T\Big|_{z=0} &= C_{T,\text{feed}} \;@ t \le t_{\text{injection}} \\ \frac{\partial C_T}{\partial z}\Big|_{z=L} &= 0 \end{aligned} \tag{5.15}$$

- *Open–open vessels*: dispersion also at the entrance. Continuity condition at the outlet, we have

$$-D_z\frac{\partial C_T}{\partial z}\bigg|_{z=0^-}+uC_T\bigg|_{z=0^-}=-D_z\frac{\partial C_T}{\partial z}\bigg|_{z=0^+}+uC_T\bigg|_{z=0^+}\quad @t\le t_{\text{injection}}\quad \frac{\partial C_T}{\partial z}\bigg|_{z=L}=0 \quad (5.16)$$

Obviously, the situation is identical in the case when a chemical reaction is also considered [3].

In MATLAB, all the mentioned cases in this paragraph can be solved using the *pdepe* function. Several examples are given as follows to illustrate how to solve different problems that an expert in the field could frequently face with.

Example 5.2 Axial dispersion, stepwise RTD experiment
For an RTD stepwise experiment, solve the axial dispersion dimensionless model, for Pe = 0.1,1,10,100,1000:

$$\frac{\partial y}{\partial \theta}=-\frac{\partial y}{\partial x}+\frac{1}{\text{Pe}}\frac{\partial^2 y}{\partial x^2}$$

Consider the pipe empty of the tracer at $t=0$.

Matlab code for the solution of this example and results are reported as follows:

```
% example 5.2
clc,clear

%% Reactor properties
X=1;
Pek=[0.1 1 10 100 1000];
xmesh=linspace(0,X,300);
tauspan=linspace(0,2,1000);

%% pdepe call
for k=1:5
    Pe=Pek(k);
    m=0;
    options=odeset('RelTol',1e-8,'AbsTol',1e-8);
    f1=@(x,t,u,dudx)pdefun5_02(x,t,u,dudx,Pe);
    f2=@(xl,ul,xr,ur,t)bcfun5_02(xl,ul,xr,ur,t);
    f3=@(x)icfun5_02(x);
    c(:,:,k)=pdepe(m,f1,f3,f2,xmesh,tauspan,options);
end

%% Output
figure(1)
subplot(1,3,1)
```

```
surf(xmesh,tauspan,c(:,:,1),'edgecolor','none')
xlabel('\it \chi \rm [-]')
ylabel('\it \theta \rm [-]')
zlabel('\it y \rm [-]')
title('\it Pe \rm = 0.1')

subplot(1,3,2)
surf(xmesh,tauspan,c(:,:,4),'edgecolor','none')
xlabel('\it \chi \rm [-]')
ylabel('\it \theta \rm [-]')
zlabel('\it y \rm [-]')
title('\it Pe \rm = 100')

subplot(1,3,3)
plot(tauspan,c(:,end,1),'-k',tauspan,c(:,end,2),'-r',tauspan,
c(:,end,3),'-c',tauspan,c(:,end,4),'-m',tauspan,c(:,end,5),'-b')
legend('\it Pe\rm = 0.1','\it Pe\rm = 1','\it Pe\rm = 10','\it Pe\rm
= 100','\it Pe\rm = 1000')
xlabel('\it \theta \rm [-]')
ylabel('\it y \rm |_{\it \chi \rm = \rm 1} [-]')
axis([0 2 0 1.05])
grid on

function [c,f,s] = pdefun5_02(x,t,u,dudx,Pe)
c=1;
f=(1/Pe)*dudx;
s=-dudx;

end

function [pl,ql,pr,qr] = bcfun5_02(xl,ul,xr,ur,t)
pl = ul-1;
ql = 0;
pr = 0;
qr = 1;
end
function u0 = icfun5_02(x)
u0=0;
end
```

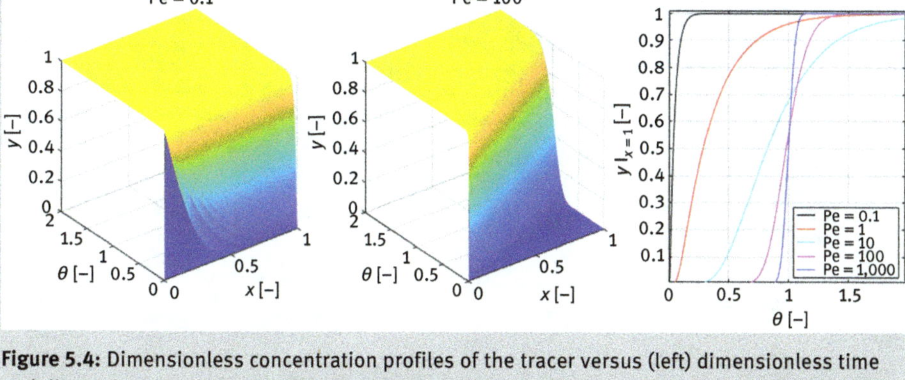

Figure 5.4: Dimensionless concentration profiles of the tracer versus (left) dimensionless time and dimensionless axial coordinate for Pe = 0.1; (center) dimensionless time and dimensionless axial coordinate for Pe = 1000; (right) dimensionless time at the outlet of the pipe parametric with the Péclet number.

As expected, higher Péclet number leads to less backmixing; thus, a fluid dynamics that approaches the plug-flow behavior (Figure 5.4).

Example 5.3 Axial dispersion, pulse RTD experiment
For an RTD pulse experiment, solve the axial dispersion dimensionless model, for Pe = 100:

$$\frac{\partial y}{\partial \theta} = -\frac{\partial y}{\partial \chi} + \frac{1}{\text{Pe}}\frac{\partial^2 y}{\partial \chi^2}$$

Consider the pipe empty of the tracer at $t = 0$. The injection time was set to 0.02.

Be careful! In this example, due to the numerical difficulty of the problem, a logarithmic spacing of the *xmesh* variable was adopted. In this way, more calculation points are set in the initial part of the pipe where the peak is sharper, approaching a δ-Dirac function.

Matlab code for the solution of this example and results are reported as follows:

```
% example 5.3
clc,clear

%% Reactor properties
X=1;

Pe=100;
tinj=0.02;

xmesh=logspace(-4,X,500)/10;
tauspan=linspace(0,2,1000);
```

```
%% pdepe call
m=0;
options=odeset('RelTol',1e-1,'AbsTol',1e-4);
f1=@(x,t,u,dudx)pdefun5_03(x,t,u,dudx,Pe);
f2=@(xl,ul,xr,ur,t)bcfun5_03(xl,ul,xr,ur,t,tinj);
f3=@(x)icfun5_03(x);
c=pdepe(m,f1,f3,f2,xmesh,tauspan,options);

%% Output
figure(1)
subplot(1,2,1)
surf(xmesh,tauspan,c,'edgecolor','none')
xlabel('\it \chi \rm [-]')
ylabel('\it \theta \rm [-]')
zlabel('\it y \rm [-]')
axis([0 1 0 2 0 1])
subplot(1,2,2)
plot(tauspan,c(:,end),'-k')
xlabel('\it \theta \rm [-]')
ylabel('\it y \rm |_{\it \chi \rm = \rm 1} [-]')
grid on
axis([0 2 0 0.07])

function [c,f,s] = pdefun5_03(x,t,u,dudx,Pe)
c=1;
f=(1/Pe)*dudx;
s=-dudx;

end

function [pl,ql,pr,qr] = bcfun(xl,ul,xr,ur,t,tinj)
if t ≤ tinj
  pl = ul-1;
else
  pl = ul ;
end

ql = 0;
pr = 0;
qr = 1;

end

function u0 = icfun5_03(x)
u0=0;

end
```

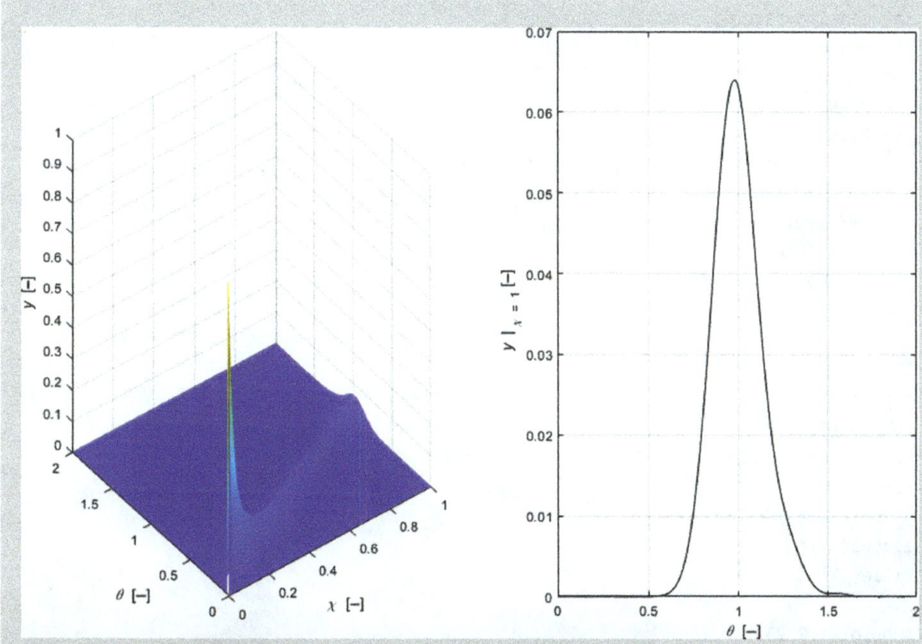

Figure 5.5: Dimensionless concentration profiles of the tracer versus (left) dimensionless time and dimensionless axial coordinate; (right) dimensionless time at the outlet of the pipe.

As revealed, along the length and time the peak width increases as Péclet number is rather low. Some numerical fluctuations are observed due to the difficulty of the solver to face with stiff problems (Figure 5.5).

Example 5.4 Axial dispersion, single reaction
For a kinetic experiment conducted in a continuous reactor, obeying to a first-order reaction kinetics, solve the axial dispersion dimensionless model, using Pe = 10. Vary the $D_A = 1,10,100$:

$$\frac{\partial y}{\partial \Theta} = -\frac{\partial y}{\partial \chi} + \frac{1}{\text{Pe}}\frac{\partial^2 y}{\partial \chi^2} - D_A y$$

Consider the pipe empty of the tracer at $t = 0$.

Matlab code for the solution of this example and results are reported as follows:

```
% example 5.4
clc,clear


%% Reactor properties
X=1;
Pe=10;
DAk=[0.01,0.1,1];
xmesh=linspace(0,X,300);
```

```
tauspan=linspace(0,5,1000);
%% pdepe call
for k=1:3
  DA=DAk(k);
  m=0;
  options=odeset('RelTol',1e-8,'AbsTol',1e-8);
  f1=@(x,t,u,dudx)pdefun5_04(x,t,u,dudx,Pe,DA);
  f2=@(xl,ul,xr,ur,t)bcfun5_04(xl,ul,xr,ur,t);
  f3=@(x)icfun5_04(x);
  c(:,:,k)=pdepe(m,f1,f3,f2,xmesh,tauspan,options);
end

%% Output
figure(1)
subplot(1,2,1)
surf(xmesh,tauspan,c(:,:,3),'edgecolor','none')
xlabel('\it \chi \rm [-]')
ylabel('\it \theta \rm [-]')
zlabel('\it y \rm [-]')
title('\it D_{A} \rm = 1')

subplot(1,2,2)
plot(tauspan,c(:,end,1),'-k',tauspan,c(:,end,2),'-r',tauspan,c(:,end,3),'-c')
legend('\it D_{A}\rm = 0.01','\it D_{A}\rm = 0.1','\it D_{A}\rm = 1')
xlabel('\it \theta \rm [-]')
ylabel('\it y \rm |_{\it \chi \rm = \rm 1} [-]')
axis([0 5 0 1.05])
grid on

function [c,f,s] = pdefun5_04(x,t,u,dudx,Pe,DA)
c=1;
f=(1/Pe)*dudx;
s=-dudx-DA*u;

end

function [pl,ql,pr,qr] = bcfun5_04(xl,ul,xr,ur,t)
pl = ul-1;
ql = 0;
pr = 0;
qr = 1;

end

function u0 = icfun5_04(x)
u0=0;

end
```

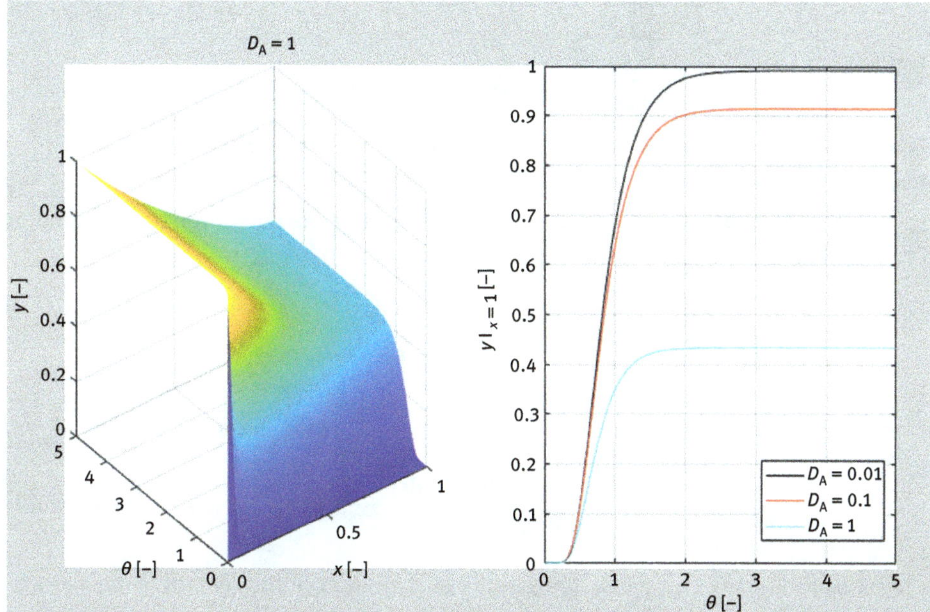

Figure 5.6: Dimensionless concentration profiles of the reactant versus (left) dimensionless time and dimensionless axial coordinate for $D_A = 1$; (right) dimensionless time at the outlet of the reactor parametric with D_A.

As expected, higher D_A lead to higher conversion. The shape of the curve resembles what it was obtained in Example 5.2, as Péclet number is the same. The plateau value, instead, strictly depends on D_A: if no reaction occurs, $y = 1$ is obtained (as for RTD tests); for high value of D_A, the plateau value is lower than 1, corresponding to the conversion of the system (Figure 5.6).

5.4 Laminar flow model

In an LFR, the axial velocity varies in the radial direction, drawing a parabolic trend (Figure 5.7) [4].

The fluid dynamics is commonly described according to the Hagen–Poiseuille equation, where the flow velocity in the middle of the pipe is two times the average fluid velocity given by the pumping system:

$$u(r) = 2\bar{u}\left[1 - \left(\frac{r}{R}\right)^2\right] \tag{5.17}$$

This equation clearly states that the velocity of the flow is zero at the pipe wall.

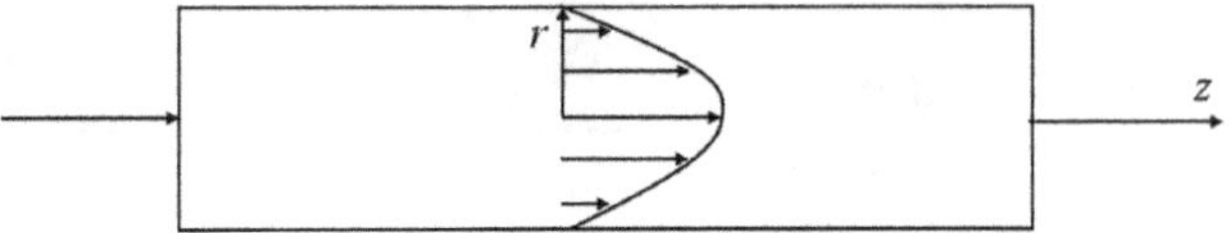

Figure 5.7: Laminar flow reactor model sketch: representation of the streamlines.

For a continuous reactor where the fluid dynamics is characterized by laminar flow, the RTD is defined as

$$E(t) = \begin{cases} 0 \ \text{for}\, t < \frac{\tau}{2} \\ \frac{\tau^2}{2t^3} \ \text{for}\ t \geq \frac{\tau}{2} \end{cases} \tag{5.18}$$

If radial diffusion occurs, it is possible to write the following mass balance equation:

$$\frac{\partial C_i}{\partial t} = -u(r)\frac{\partial C_i}{\partial z} + D_i\left[\frac{1}{r}\frac{\partial r(\partial C_i/\partial r)}{\partial r} + \frac{\partial^2 C_i}{\partial z^2}\right] + \sum_k \nu_{ik} r_k \tag{5.19}$$

It must be reminded that what is measured at the outlet of the reactor is the average concentration along the radial coordinate, which could be calculated in agreement with the following equation:

$$\bar{C}_i(z,t) = \frac{1}{\pi R^2}\int_0^R C_i(r,z,t) 2\pi r dr \tag{5.20}$$

If the flow is fully laminar, no axial diffusion is present; thus, the mass balance equation is simplified:

$$\frac{\partial C_i}{\partial t} = -u(r)\frac{\partial C_i}{\partial z} + D_i\left[\frac{1}{r}\frac{\partial r(\partial C_i/\partial r)}{\partial r}\right] + \sum_k \nu_{ik} r_k \tag{5.21}$$

As revealed, the concentration of component i varies both with reaction time, the axial position of the reactor by convection and the radial position of the pipe by molecular diffusion. Solving such a problem implies the numerical implementation of a three-dimension problem. As most of the cases, the researchers are interested at the steady-state solution, and the problem can be rewritten deleting the accumulation term:

$$u(r)\frac{\partial C_i}{\partial z} = +D_i\left[\frac{1}{r}\frac{\partial r(\partial C_i/\partial r)}{\partial r}\right] + \sum_k \nu_{ik} r_k \tag{5.22}$$

To solve this Partial Differential Equation (PDE) system, it is necessary to remind that the velocity must be calculated at each radial position at each axial coordinate. Moreover, two boundary conditions are needed:

- Center of the reactor: symmetry condition due to the parabolic shape of the velocity pattern profile

$$\left.\frac{\partial C_i}{\partial r}\right|_{r=0} = 0 \tag{5.23}$$

- Reactor wall: nonslip condition

$$\left.\frac{\partial C_i}{\partial r}\right|_{r=R} = 0 \tag{5.24}$$

In this form, the equation can be solved using the *pdepe* function already implemented in MATLAB, as the axial coordinate is solved automatically by *ode15s* function. An example of implementation is provided as follows.

Example 5.5 Laminar flow model
Solve a laminar flow model for a first-order reaction rate, adopting the mass balance reported as follows:

$$u(r)\frac{\partial C_A}{\partial z} = +D_A\left[\frac{1}{r}\frac{\partial r(\partial C_A/\partial r)}{\partial r}\right] - kC_A$$

The settings needed are reported in Table 5.1.

Table 5.1: Parameters needed to solve the laminar flow reactor model.

Parameter	Value	Units
$C_{A,\text{feed}}$	1,000	mol/m^3
L	0.5	m
R	1e-2	m
$\bar{u}$	1e-5	m/s
D_i	1e-9	m^2/s
k	1e-4	1/s

Consider the pipe empty of the tracer at $t = 0$.

Matlab code for the solution of this example and results are reported as follows:

```
% example 5.5
clc,clear

%% Reactor properties
CAfeed=1000;
L=0.5;
R=1e-2;
uav=1e-5;
Di=1e-9;
k=1e-4;
rmesh=linspace(0,R,300);
zspan=linspace(0,L,1000);

%% pdepe call
m=1;
options=odeset('RelTol',1e-8,'AbsTol',1e-8);
f1=@(x,t,u,dudx)pdefun5_05(x,t,u,dudx,k,uav,Di,R);
f2=@(xl,ul,xr,ur,t)bcfun5_05(xl,ul,xr,ur,t);
f3=@(x)icfun5_05(x,CAfeed);
c=pdepe(m,f1,f3,f2,rmesh,zspan,options);

%% Output
figure(1)
subplot(1,3,1)
surf(rmesh,zspan,c,'edgecolor','none')
xlabel('\it r \rm [m]')
ylabel('\it z \rm [m]')
zlabel('\it C_{A} \rm [mol/m^{3}]')

subplot(1,3,2)
plot(zspan,c(:,1),'-b',zspan,c(:,end/2),'-r',zspan,c(:,end),'-k')
xlabel('\it z \rm [m]')
ylabel('\it C_{A} \rm [mol/m^{3}]')
grid on
legend('\it r\rm = \it 0','\it r\rm = 0.5\it R\rm','\it r\rm = \it R')

subplot(1,3,3)
plot(rmesh,c(end,:))
xlabel('\it r \rm [m]')
ylabel('\it C_{A} \rm |_{\it z \rm = \rm \it L \rm} [mol/m^{3}]')
grid on

function [c,f,s] = pdefun5_05(x,t,u,dudx,k,uav,Di,R)
c=2*uav*(1-(x/R)^2);
f=Di*dudx;
s=-k*u;
end
```

```
function [pl,ql,pr,qr] = bcfun5_05(xl,ul,xr,ur,t)
pl = 0;
ql = 1;
pr = 0;
qr = 1;
end
function u0 = icfun5_05(x,CAfeed)
u0=CAfeed;
end
```

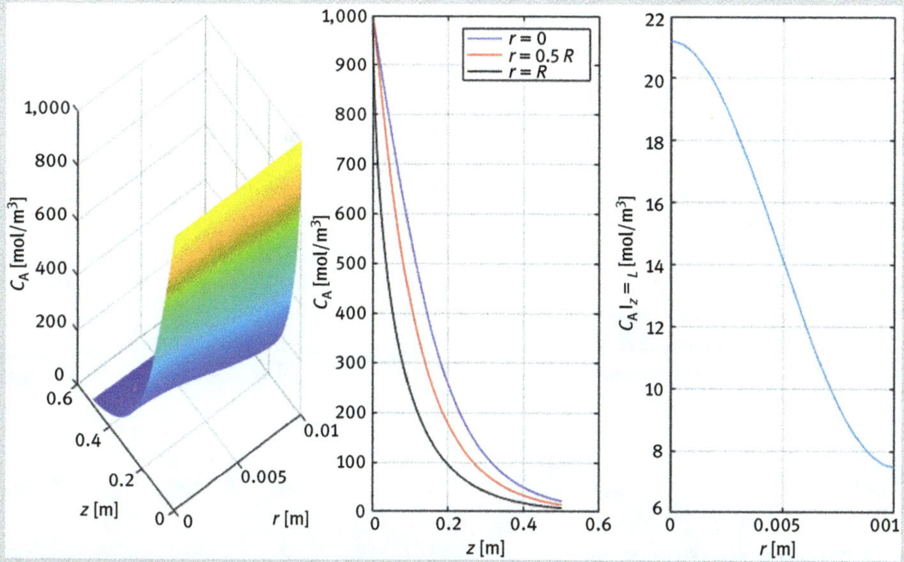

Figure 5.8: Concentration profiles of the reactant versus both axial and radial coordinate of the reactor (left) ; (center) concentration profile along the axial coordinate parametric with the radial position; (right) concentration of reactant A along the radial coordinate at the outlet of the reactor.

The concentration profile of component A shows a parabolic trend along the radial coordinate as expected (Figure 5.8). From the parametric plot, it is evident that at the center of the pipe we get the lowest conversion, as the velocity of the flow is the highest that lead to the lowest residence time, while it is maximum at the pipe walls. Finally, cutting the surface at the outlet of the pipe, it is evident that the concentration of component A profile shows a maximum at the center of the pipe.

List of symbols

A	Tubular reactor section	[m^2]
C_i	Concentration of component i	[mol/m^3]
$\bar{C}_i$	Average concentration of component i	[mol/m^3]
C_T	Concentration of the tracer	[mol/m^3]
D_A	Damkohler number	[–]
D_i	Molecular diffusivity	[m^2/s]
D_z	Axial dispersion coefficient	[m^2/s]
E	Residence time distribution function	[–]
L	Pipe length	[m]
n	Number of CSTR	[–]
n_i	Amount of substance	[mol]
$\dot{n}_i$	Molar flowrate	[mol/s]
r	Pipe radial coordinate	[m]
r_j	Reaction rate	[mol/(m^3 s)]
R	Pipe radius	[m]
Pe	Péclet number	[–]
t	Time	[s]
u	Fluid velocity	[m/s]
V	Volume	[m^3]
$\dot{V}$	Volumetric flowrate	[m^3/s]
r	Dimensionless concentration	[–]
z	Pipe axial coordinate	[m]

Greek letters

χ	Dimensionless axial coordinate	[–]
Θ	Dimensionless time	[–]
τ	Residence time	[s]
ν_{ik}	Stoichiometric coefficient of component i for reaction k	[–]

References

[1] O. Levenspiel. Chemical Reaction Engineering (3rd edition). Wiley: 1998.

[2] H.S. Fogler. Elements of Chemical Reaction Engineering (5th edition). Prentice Hall: 2016.

[3] V. Russo, T. Salmi, F. Mammitzsch, O. Jogunola, R. Lange, J. Wärnåa, J.-P. Mikkola. First, second and nth order autocatalytic kinetics in continuous and discontinuous reactors. Chemical Engineering Science 2017, 172, 453–462.

[4] A.P. Nebreda, V. Russo, M. Di Serio, T. Salmi, H. Grénman. Modelling of homogeneously catalyzed hemicelluloses hydrolysis in a laminar-flow reactor. Chemical Engineering Science 2019, 195, 758–766.

Chapter 6
Packed bed reactors

6.1 Introduction

Packed bed reactors (PBR) can be considered among the most common devices to conduct experiments in continuous testing heterogeneous catalysts. It is a common practice to conduct the experiments in order to exclude, or at least minimize, all the phenomena that could avoid the occurrence of kinetic regime. Working at high flow velocity and low temperature with small-sized catalyst particle is certainly a good lab practice; therefore, due to intrinsic characteristics of the chemical system under investigation, sometimes it is rather difficult to keep the system ideal. Very often, axial dispersion, fluid–solid mass transfer or intraparticle mass transfer limitations are present [1–3]. Thus, the reactor model should be simple enough but rigorous to describe the main phenomena occurring. In this chapter, the main PBR models are shown with coding examples.

6.2 Ideal packed bed reactor

The ideal PBR model is maybe the most used reactor model for this category. Talking about ideality means that both fluid–solid and intraparticle diffusion limitation can be considered negligible. Moreover, the fluid dynamic is described simply by a plug-flow approach. To describe the system, the molar balance for the ith component is written by choosing as integration volume a slice of infinitesimal depth of either volume/length of the packed pipe or catalyst mass (see sketch in Figure 6.1).

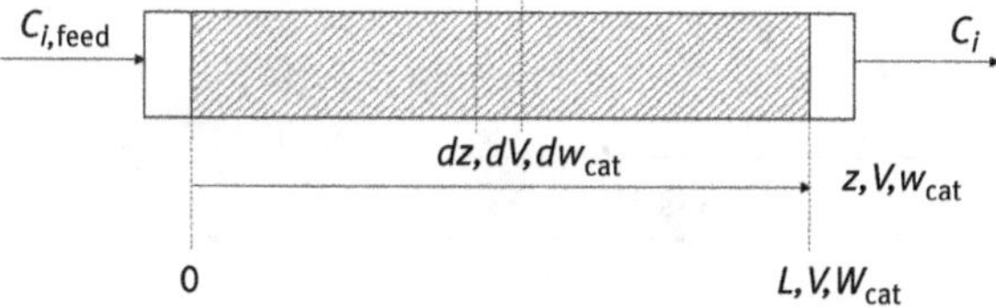

Figure 6.1: Ideal PBR modeling sketch.

Both mass and energy balance equations can be derived from the general conservation equation that can be written as follows:

$$[\text{inlet}_i] - [\text{outlet}_i] = [\text{generation}_i] \tag{6.1}$$

By introducing the correct terms, it is possible to obtain the following mass and energy balances:

https://doi.org/10.1515/9783110632927-006

$$u\frac{dC_i}{dz} = \sum_{k=1}^{N_r} \nu_{ik} r_k \rho_B, \quad u = \frac{\dot{V}}{A\varepsilon_B} \tag{6.2}$$

$$\left(\sum_{i=1}^{N_C} C_i c_{p,i}\right) u\frac{dT}{dz} = \sum_{k=1}^{N_r} (-\Delta_r H_k) r_k \rho_B - U\frac{A}{V}(T - T_J) \tag{6.3}$$

C_i is the concentration of the ith component in the bulk phase, u the linear fluid velocity [m/s] obtained by the volumetric flowrate and the free section of the pipe, r is the reaction rate given for catalyst mass units [mol/(kg s)], ρ_B is catalyst bulk density, $c_{p,i}$ the molar specific heat of each component, $\Delta_r H$ the reaction heat [J/mol]. Very commonly, the heat balance equation is coupled with a heat exchange term against a constant temperature of the heating/cooling jacket (T_J). Obviously, in this case it is necessary to know a priori the overall heat exchange coefficient U [J/(m K s)], obtainable with dedicated experiments.

As revealed, the two equations are ordinary differential equations; thus, it is possible to solve them in MATLAB by using one of the *ode* algorithms. Two examples are provided in this textbook, to show the readers how to solve both an isothermal and heat exchanged PBR, this last in the case of a constant jacket temperature.

Example 6.1 Ideal isothermal PBR

For a given first-order reaction A → B, solve the ideal isothermal PBR model. The settings needed are listed in Table 6.1.

Table 6.1: Settings needed to solve the simplified intraparticle PBR model.

Parameter	Value	Units
U	1e-4	m/s
C_{feed}	1e3	mol/m^3
T	340	K
L	1	m
k_{ref}	1e-6	m^3/(kg s)
T_{ref}	340	K
E_a	120	kJ/mol
ρ_B	1,500	kg/m^3

Matlab code for the solution of this example and results are reported as follows Figure 6.2:

```
% example 6.1
clc,clear

%% Reactor Settings
u = 1e-4 ;              % m/s
rhoB = 1500 ;           % kg/m3
Cfeed = 1e3 ;           % mol/m3
T = 340 ;               % K
L = 1 ;                 % m

%% Kinetics Settings
kref=1e-6;              % m3/(kg s)
Tref=340;               % K
Ea=120000;              % J/mol
R=8.3144;               % J/(K mol)

%% ode call
zspan=linspace(0,L,1e2);
f1=@(z,V)odefun6_01(z,V,kref,Ea,R,Tref,u,rhoB,T);
[z,V]=ode15s(f1,zspan,Cfeed);
ca=V(:,1);

%% Figures
plot(zspan,ca,'-r')
xlabel('\it z \rm [m]')
ylabel('\it C_{A} \rm [mol/m^{3}]')
grid on
function [dVdz] = odefun6_01(z,V,kref,Ea,R,Tref,u,rhoB,T)
ca=V(1);

r=kinet_iso6_01(ca,T,kref,Tref,Ea,R); % mol/(kg s)

%% Bulk phase
dVdz(1)=-r*rhoB/u;

dVdz=dVdz';
end

function [ reaz ] = kinet_iso6_01(ca,T,kref,Tref,Ea,R)
k=kref*exp(-Ea/R*(1/T-1/Tref));
reaz=k.*ca;
end
```

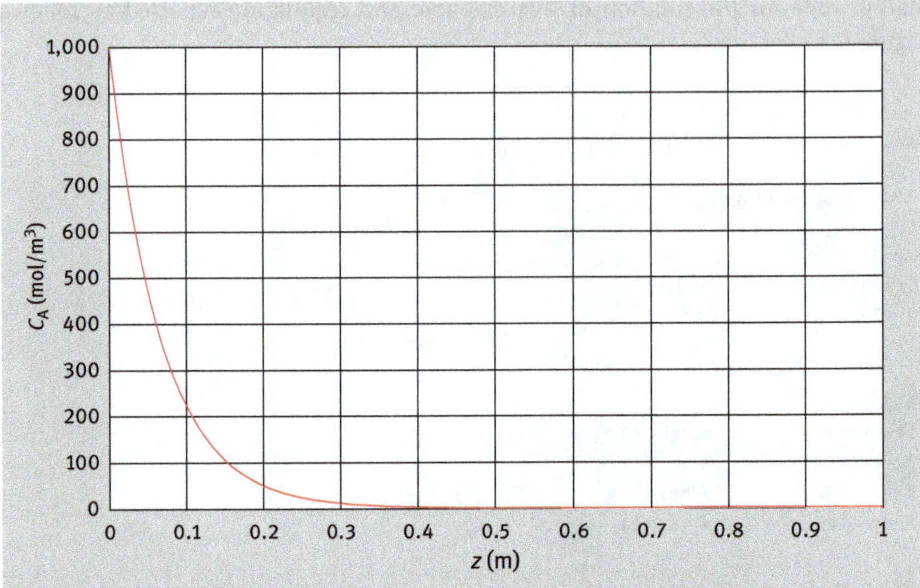

Figure 6.2: Concentration profiles as a function of the axial coordinate of the reactor.

Conversion increases with the reactor length coordinate as expected, reaching full conversion at the outlet of the pipe.

Example 6.2 Heat exchanged PBR

For a given first-order exothermic reaction A → B, solve the ideal nonisothermal PBR model. The settings needed are listed in Table 6.2.

Table 6.2: Settings needed to solve the simplified intraparticle PBR model.

Parameter	Value	Units
u	1e-4	m/s
C_{feed}	1e3	mol/m^3
T_{feed}	340	K
L	1	m
UA/V	1,000	J/(m^2 K s)
T_J	340	K
$c_{pA} = c_{pB}$	70	J/(mol K)
k_{ref}	1e-6	m^3/(kg s)

Table 6.2 (continued)

Parameter	Value	Units
T_{ref}	340	K
E_a	120	kJ/mol
$\Delta_r H$	−25	kJ/mol
ρ_B	1,500	kg/m^3

Matlab code for the solution of this example and results are reported below Figure 6.3.

```
% example 6.2
clc,clear

%% Reactor Settings
u = 1e-4 ;           % m/s
rhoB = 1500 ;        % kg/m3
Cfeed = 1e3 ;        % mol/m3
Tfeed = 340 ;        % K
L = 1 ;              % m
UAV = 1000 ;         % J/(m2 K s)
TJ = 340 ;           % K
cpA = 70 ;           % J/(mol K)
cpB = 70 ;           % J/(mol K)

%% Kinetics Settings
kref=1e-7;            % (m3/kg)(m3/mol)(1/s)
Tref=340;             % K
Ea=120000;            % J/mol
R=8.3144;             % J/(K mol)
DrH = -25000 ;        % J/mol

%% ode call
Vfeed=[Cfeed,Tfeed];
zspan=linspace(0,L,1e2);
f1=@(z,V)odefun6_02(z,V,kref,Ea,R,Tref,u,rhoB,DrH,UAV,TJ,cpA,cpB,Cfeed);
[z,V]=ode15s(f1,zspan,Vfeed);
ca=V(:,1);
T=V(:,2);

%% Figures
subplot(1,2,1)
plot(zspan,ca,'-r')
xlabel('\it z \rm [m]')
ylabel('\it C_{A} \rm [mol/m^{3}]')
grid on
subplot(1,2,2)
```

```
plot(zspan,T,'-r')
xlabel('\it z \rm [m]')
ylabel('\it T \rm [K]')
grid on

function [dVdz] = odefun6_02(z,V,kref,Ea,R,Tref,u,rhoB,DrH,UAV,TJ,cpA,cpB,Cfeed)
ca=V(1);
T=V(2);

r=kinetic6_02(ca,T,kref,Tref,Ea,R); % mol/(kg s)

A=(cpA*ca)+(cpB*(Cfeed-ca));

%% Bulk phase
dVdz(1)=-r*rhoB/u;
dVdz(2)=((-DrH).*r.*rhoB-UAV.*(T-TJ))./(u.*A);

dVdz=dVdz';
end

function [ reaz ] = kinetic6_02(ca,T,kref,Tref,Ea,R)
k=kref*exp(-Ea/R*(1/T-1/Tref));
reaz=k.*ca;

end
```

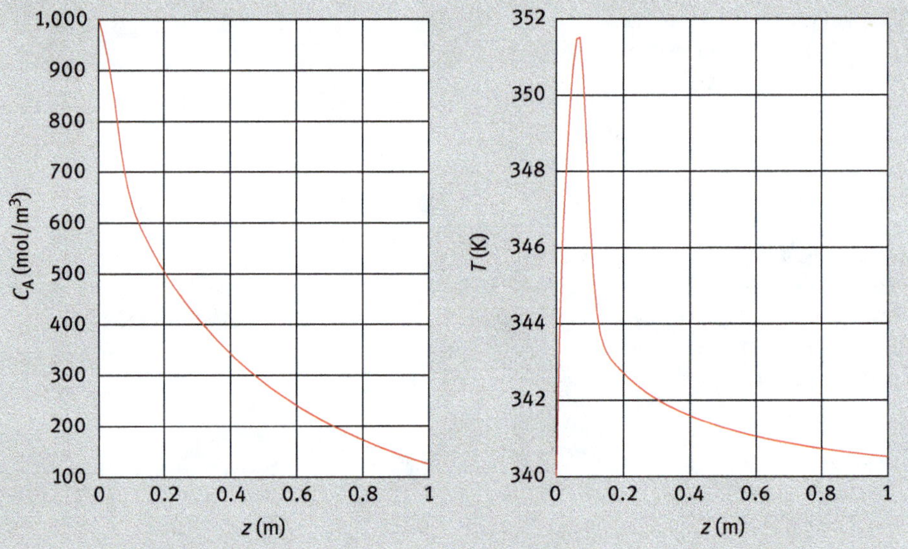

Figure 6.3: (Left) Concentration and (right) temperature profiles as a function of the axial coordinate of the reactor.

Conversion increases with the reactor length coordinate as expected, while temperature shows a maximum, due to the balance between the heat generated by the chemical reaction and the heat exchanged with the jacket.

The dynamic terms of both mass and energy balance equations are often neglected, usually when steady-state information is the needed output. Therefore, to simulate the transient state of the experiment, it is necessary to also introduce the time derivatives

$$\frac{\partial C_i}{\partial t} = -u\frac{\partial C_i}{\partial z} + \sum_{k=1}^{N_\mathrm{r}} \nu_{ik} r_k \rho_\mathrm{B} \tag{6.4}$$

$$\left(\sum_{i=1}^{N_\mathrm{C}} C_i c_{p,i}\right)\frac{\partial T}{\partial t} = -\left(\sum_{i=1}^{N_\mathrm{C}} C_i c_{p,i}\right)u\frac{\partial T}{\partial z} + \sum_{k=1}^{N_\mathrm{r}} (-\Delta_r H_k) r_k \rho_\mathrm{B} - U\frac{A}{V}(T - T_\mathrm{J}) \tag{6.5}$$

In this case, partial differential equations must be solved, thus the correct definition of the boundary conditions is required. The most common Boundary Conditions (B.C.) adopted for ideal PBR are listed as follows:

$$C_i|_{z=0} = C_{i,\,\mathrm{feed}} \tag{6.6}$$

$$\frac{\partial C_i}{\partial z}\Big|_{z=L} = 0 \tag{6.7}$$

$$T|_{z=0} = T_\mathrm{feed} \tag{6.8}$$

$$\frac{\partial T}{\partial z}\Big|_{z=L} = 0 \tag{6.9}$$

Both concentrations and temperature are fixed at a given value, as no backmixing is possible due to the ideality condition. At the outlet of the reactor, a zero derivative is assumed.

Two examples of solutions are given in this book, both solved using the *pdepe* algorithm, analyzing the cases of PBRs working either in isothermal or heat exchanged condition. In particular, the models were written including an axial dispersion term rather negligible.

Example 6.3 Dynamic PBR working in isothermal conditions

For a given first-order reaction A → B, solve the ideal dynamic PBR model. The settings needed are listed in Table 6.3.

Table 6.3: Settings needed to solve the simplified intraparticle PBR model.

Parameter	Value	Units
u	1e-3	m/s
C_{feed}	1	mol/m^3
T	340	K
L	1	m
k_{ref}	1e-6	m^3/(kg s)
T_{ref}	340	K
E_a	120	kJ/mol
ρ_B	1,500	kg/m^3

Matlab code for the solution of this example and results are reported as follows Figure 6.3:

```
% example 6.3
clc,clear

%% Reactor Settings
uf = 1e-3 ;      % m/s
rhoB = 1500 ;    % kg/m3
Cfeed = 1 ;      % mol/m3
T = 340 ;        % K
L = 1 ;          % m

zmesh=linspace(0,L,100);
tspan=linspace(0,2000,1e3);

%% Kinetics Settings
kref=1e-6;     % m3/(kg s)
Tref=340;      % K
Ea=120000;     % J/mol
R=8.3144;      % J/(K mol)

%% pdpe call
m=0;

f1=@(x,t,u,dudx)pdepefun6_03(x,t,u,dudx,uf,kref,Ea,R,Tref,rhoB,T);
f2=@(xl,ul,xr,ur,t)bcfun6_03(xl,ul,xr,ur,t,Cfeed);
f3=@(x)icfun6_03(x);
options=odeset('RelTol',1e-3,'AbsTol',1e-4);
c=pdepe(m,f1,f3,f2,zmesh,tspan,options);
```

```
%% odecall
f4=@(z,ca)odefun6_03(z,ca,uf,kref,Ea,R,Tref,rhoB,T);
[z,ca] = ode15s(f4,zmesh,Cfeed);

%% Figures
plot(tspan,c(:,end),'-r',tspan(end),ca(end),'or')
xlabel('\it t \rm [s]')
ylabel('\it C_{A} \rm|_{\it z \rm= \itL} [mol/m^{3}]')
grid on

function [c,f,s] = pdepefun6_03(x,t,u,dudx,uf,kref,Ea,R,Tref,rhoB,T)
ca=u(1);

r=kinetic6_03(ca,T,kref,Tref,Ea,R); % mol/(kg s)

c=1;
f=+1e-6*dudx;
s=-uf*dudx+rhoB*(-r);

end

function [pl,ql,pr,qr]=bcfun6_03(xl,ul,xr,ur,t,Cfeed)
pl=ul-Cfeed;
ql=0;
pr=0;
qr=1;

end

function u0=icfun6_03(x)
u0(1,1)=0;

end

function [ reaz ] = kinetic6_03(ca,T,kref,Tref,Ea,R)
k=kref*exp(-Ea/R*(1/T-1/Tref));
reaz=k.*ca;

end

function [dcadz] = odefun6_03(z,ca,uf,kref,Ea,R,Tref,rhoB,T)
r=kinetic6_03(ca,T,kref,Tref,Ea,R); % mol/(kg s)
dcadz=rhoB*(-r)/uf;

end
```

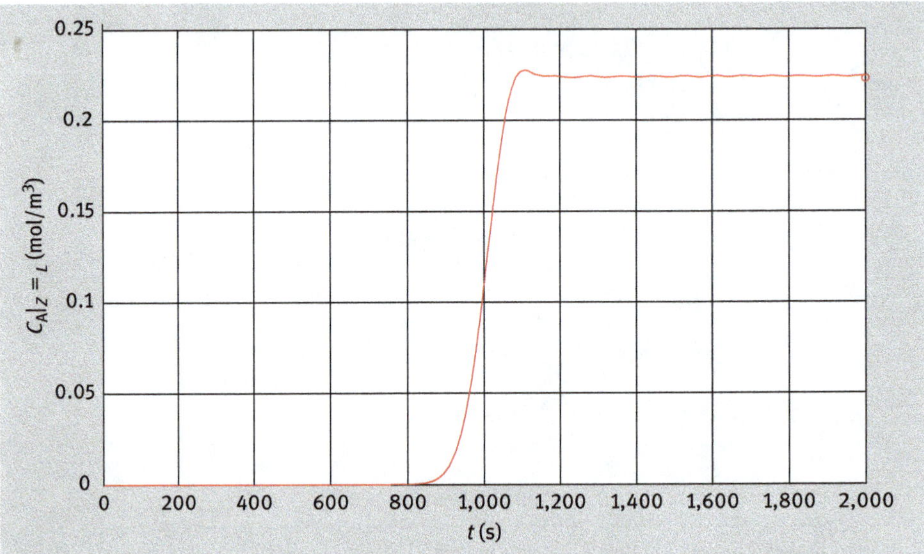

Figure 6.4: Concentration of the reactant at the outlet of the reactor as a function of time. The symbol represents the steady-state result obtained with independent model.

As the reactor is empty at $t = 0$, the concentration is zero till the average residence time, reaching then the steady-state value. The symbol represents the steady-state result obtained by using an independent ideal PBR model. As revealed, the dynamic model reaches the steady-state values. scillations are due to numerical instability due to the low dispersion coefficient adopted.

Example 6.4 Dynamic heat-exchanged PBR

For an exothermic first-order reaction, solve the ideal dynamic PBR model. The settings needed are listed in Table 6.4.

Table 6.4: Settings needed to solve the simplified intraparticle PBR model.

Parameter	Value	Units
u	1e-3	m/s
C_{feed}	1	mol/m^3
T_{feed}	340	K
L	1	m
T_J	340	K

Table 6.4 (continued)

Parameter	Value	Units
k_{ref}	1e-6	$m^3/(kg\ s)$
T_{ref}	340	K
E_a	120	kJ/mol
$\Delta_r H$	−50	kJ/mol
UA/V	50	$J/(m^2\ s\ K)$
$c_{pA} = c_{pB}$	70	J/(mol K)
ρ_B	1,500	kg/m^3

Matlab code for the solution of this example and results are reported as follows Figure 6.5 and 6.6:

```
% example 6.4
clc,clear

%% Reactor Settings
uf = 1e-3 ;      % m/s
rhoB = 1500 ;    % kg/m3
Cfeed = 1 ;      % mol/m3
Tfeed = 340 ;    % K
TJ = 340 ;       % K
L = 1 ;          % m

Vfeed = [Cfeed, Tfeed];
zmesh=linspace(0,L,100);
tspan=linspace(0,2000,1e3);

%% Kinetics Settings
kref=1e-6;       % m3/(kg s)
Tref=340;        % K
Ea=120000;       % J/mol
R=8.3144;        % J/(K mol)
DrH=-50000;      % J/mol
UAV=50;          % J/(m2 s K)
cpA=70;          % J(mol K)
cpB=70;          % J(mol K)

%% pdpe call
m=0;
```

```
f1=@(x,t,u,dudx)pdepefun6_04(x,t,u,dudx,uf,kref,Ea,R,Tref,rhoB,TJ,UAV,DrH,cpA,cpB,
Cfeed);
f2=@(xl,ul,xr,ur,t)bcfun6_04(xl,ul,xr,ur,t,Cfeed,Tfeed);
f3=@(x)icfun6_04(x,Tfeed);
options=[];%odeset('RelTol',1e-3,'AbsTol',1e-4);
c=pdepe(m,f1,f3,f2,zmesh,tspan,options);
%% odecall
f4=@(z,v)odefun6_04(z,v,uf,kref,Ea,R,Tref,rhoB,TJ,DrH,UAV,cpA,cpB,Cfeed);
[z,v] = ode15s(f4,zmesh,Vfeed);
ca=v(:,1);
T=v(:,2);

figure(1)
subplot(1,2,1)
plot(tspan,c(:,end,1),'-r',tspan(end),ca(end),'or')
xlabel('\it t \rm [s]')
ylabel('\it C_{A} \rm|_{\it z \rm= \itL} [mol/m^{3}]')
grid on

subplot(1,2,2)
plot(tspan,c(:,end,2),'-r',tspan(end),T(end),'or')
xlabel('\it t \rm [s]')
ylabel('\it T \rm|_{\it z \rm= \itL} [K]')
grid on

figure(2)
subplot(1,2,1)
surf(zmesh,tspan,c(:,:,1),'EdgeColor','none')
xlabel('\it z \rm [m]')
ylabel('\it t \rm [s]')
zlabel('\it C_{A} \rm [mol/m^{3}]')

subplot(1,2,2)
surf(zmesh,tspan,c(:,:,2),'EdgeColor','none')
xlabel('\it z \rm [m]')
ylabel('\it t \rm [s]')
zlabel('\it T \rm [K]')

function [c,f,s] = pdepefun6_04(x,t,u,dudx,uf,kref,Ea,R,Tref,rhoB,TJ,UAV,DrH,cpA,cpB,
Cfeed)
ca=u(1);
T=u(2);

r=kinetic6_04(ca,T,kref,Tref,Ea,R); % mol/(kg s)

c(1,1)=1;
f(1,1)=+1e-6*dudx(1);
```

```
s(1,1)=-uf*dudx(1)+rhoB*(-r);
A = cpA*ca+cpB*(Cfeed-ca);
c(2,1)=A;
f(2,1)=+1e-6*dudx(2);
s(2,1)=-uf*A*dudx(2)+rhoB*r*(-DrH)-UAV*(T-TJ);

end

function [pl,ql,pr,qr]=bcfun6_04(xl,ul,xr,ur,t,Cfeed,Tfeed)
pl(1,1)=ul(1)-Cfeed;
ql(1,1)=0;
pr(1,1)=0;
qr(1,1)=1;

pl(2,1)=ul(2)-Tfeed;
ql(2,1)=0;
pr(2,1)=0;
qr(2,1)=1;

end

function u0=icfun6_04(x,Tfeed)
u0(1,1)=0;
u0(2,1)=Tfeed;

end

function [ reaz ] = kinetic6_04(ca,T,kref,Tref,Ea,R)
k=kref*exp(-Ea/R*(1/T-1/Tref));
reaz=k.*ca;

end

function [dvdz] = odefun6_04(z,v,uf,kref,Ea,R,Tref,rhoB,TJ,DrH,UAV,cpA,cpB,Cfeed)
ca=v(1);
T=v(2);

r=kinetic6_04(ca,T,kref,Tref,Ea,R); % mol/(kg s)

dvdz(1)=rhoB*(-r)/uf;

A=cpA*ca+cpB*(Cfeed-ca);
dvdz(2)=(rhoB*r*(-DrH)-UAV*(T-TJ))/(uf*A);

dvdz=dvdz';
end
```

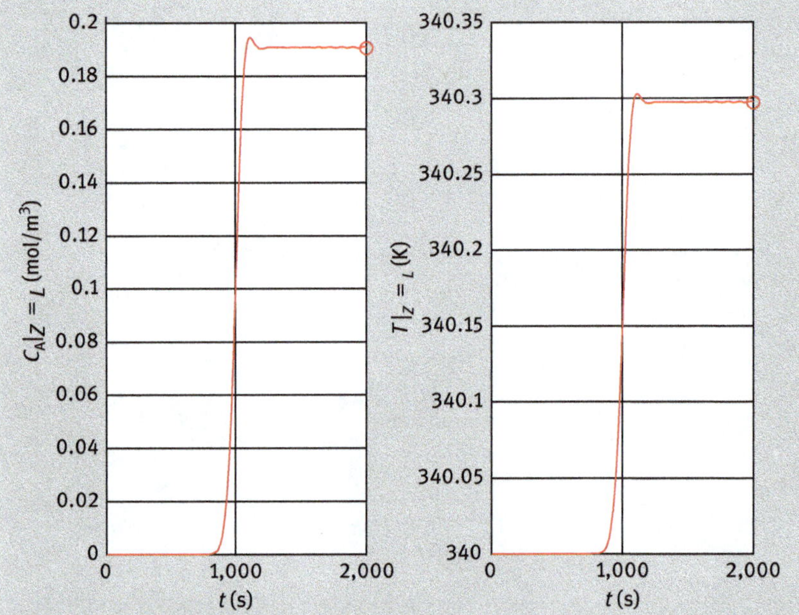

Figure 6.5: Concentration of the reactant (left) and temperature (right) profiles at steady state for both dynamic and steady-state models.

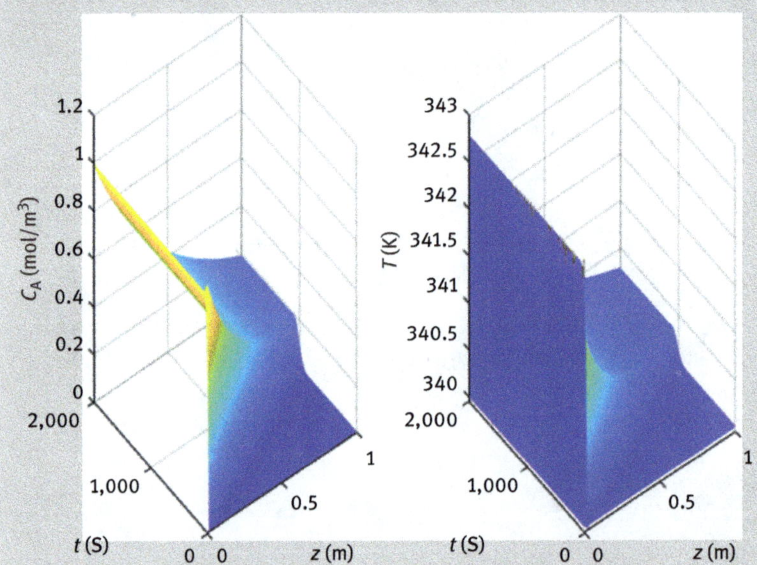

Figure 6.6: Concentration of the reactant (left) and temperature (right) surface plots versus time and reactor axial coordinates.

Also in this case, the transient model reaches the steady-state value given by the classical ideal PBR steady-state model. The surface plots show reasonable trends for both A concentration and fluid temperature, this last showing a maximum as expected.

6.3 Packed bed with axial and radial dispersion

The first nonideality that we are going to introduce is the occurrence of backmixing, that is, a generally always present phenomenon for both lab-scale and pilot-plant PBR. The packing material of a PBR, tha could be either a catalyst or a mixture of catalyst and inert solid, deviates the flow pattern of the fluid phase. This phenomenon inevitably leads to the formation of eddies that generate local turbulence, leading to averagely high Péclet numbers even if at low flowrates, required to ensure enough residence time for the reaction. Even though, the fluid dynamics can be considered far to be ideal, thus an axial dispersion term must be added both in the bass and energy balance equations, becoming

$$\frac{\partial C_i}{\partial t} = -u\frac{\partial C_i}{\partial z} + D_z\frac{\partial^2 C_i}{\partial z^2} + \sum_{k=1}^{N_r} \nu_{ik} r_k \rho_B \tag{6.10}$$

$$\left(\sum_{i=1}^{N_C} C_i c_{p,i}\right)\frac{\partial T}{\partial t} = -\left(\sum_{i=1}^{N_C} C_i c_{p,i}\right) u\frac{\partial T}{\partial z} + \lambda_z\frac{\partial^2 T}{\partial z^2} + \sum_{k=1}^{N_r}(-\Delta_r H_k) r_k \rho_B - U\frac{A}{V}(T - T_J) \tag{6.11}$$

With D_z the axial dispersion coefficient [m^2/s] and λ_z the bed axial conductivity [J/(s m K)]. Equations (6.6)–(6.9) could be used as boundary conditions for relatively high Péclet numbers. In the case of very low flowrate, thus high axial dispersion, the authors suggest the use of open–open vessel boundary conditions.

Example 6.5 shows the implementation of an axial dispersion PBR considering an isothermal case. As it will be seen, the system was solved by using the *pdepe* algorithm.

Example 6.5 Axial dispersion dynamic isothermal PBR
For a given first-order reaction A → B, solve the ideal dynamic PBR model for different values of the axial dispersion coefficient. The settings needed are listed in Table 6.5.

Table 6.5: Settings needed to solve the simplified intraparticle PBR model.

Parameter	Value	Units
u	1e-3	m/s
C_{feed}	1	mol/m^3

Table 6.5 (continued)

Parameter	Value	Units
T	340	K
L	1	m
k_{ref}	1e-6	$m^3/(kg\ s)$
T_{ref}	340	K
E_a	120	kJ/mol
ρ_B	1,500	kg/m^3
D_z	[1e-1,1e-2,1e-3,1e-4,1e-5]	m^2/s

Matlab code for the solution of this example and results are reported as follows Figure 6.7:

```
% example 6.5
clc,clear

%% Reactor Settings
uf = 1e-3 ;                       % m/s
rhoB = 1500 ;                     % kg/m3
Cfeed = 1 ;                       % mol/m3
T = 340 ;                         % K
L = 1 ;                           % m
Dz = [1e-1 1e-2 1e-3 1e-4 1e-5];  % m2/s

zmesh=linspace(0,L,100);
tspan=linspace(0,2000,1e3);

%% Kinetics Settings
kref=1e-6;                        % m3/(kg s)
Tref=340;                         % K
Ea=120000;                        % J/mol
R=8.3144;                         % J/(K mol)

%% pdpe call
m=0;

for i=1:length(Dz)
    Dzi=Dz(i);

    f1=@(x,t,u,dudx)pdepefun6_05(x,t,u,dudx,uf,kref,Ea,R,Tref,rhoB,T,Dzi);
    f2=@(xl,ul,xr,ur,t)bcfun6_05(xl,ul,xr,ur,t,Cfeed);
    f3=@(x)icfun6_05(x);
    options=odeset('RelTol',1e-3,'AbsTol',1e-4);
    c=pdepe(m,f1,f3,f2,zmesh,tspan,options);
```

```
    cx{i}=c;
end

%% odecall
f4=@(z,ca)odefun6_05(z,ca,uf,kref,Ea,R,Tref,rhoB,T);
[z,ca] = ode15s(f4,zmesh,Cfeed);

%% Figures
plot(tspan,cx{1}(:,end),tspan,cx{2}(:,end),tspan,cx{3}(:,end),tspan,cx{4}(:,end),
tspan,cx{5}(:,end), . . .
tspan(end),ca(end),'ok')
xlabel('\it t \rm [s]')
ylabel('\it C_{A} \rm|_{\it z \rm= \itL} [mol/m^{3}]')
legend('\it D\rm_{z}=1e-1','\it D\rm_{z}=1e-2','\it D\rm_{z}=1e-3','\it D\rm_{z}=1e-
4','\it D\rm_{z}=1e-5')

grid on

function [c,f,s] = pdepefun6_05(x,t,u,dudx,uf,kref,Ea,R,Tref,rhoB,T,Dz)
ca=u(1);

r=kinetic6_05(ca,T,kref,Tref,Ea,R); % mol/(kg s)

c=1;
f=+Dz*dudx;
s=-uf*dudx+rhoB*(-r);
end

function [pl,ql,pr,qr]=bcfun6_05(xl,ul,xr,ur,t,Cfeed)
pl=ul-Cfeed;
ql=0;
pr=0;
qr=1;
end

function u0=icfun6_05(x)
u0(1,1)=0;
end

function [ reaz ] = kinetic6_05(ca,T,kref,Tref,Ea,R)
k=kref*exp(-Ea/R*(1/T-1/Tref));
reaz=k.*ca;
end

function [dcadz] = odefun6_05(z,ca,uf,kref,Ea,R,Tref,rhoB,T)
r=kinetic6_05(ca,T,kref,Tref,Ea,R); % mol/(kg s)
dcadz=rhoB*(-r)/uf;
end
```

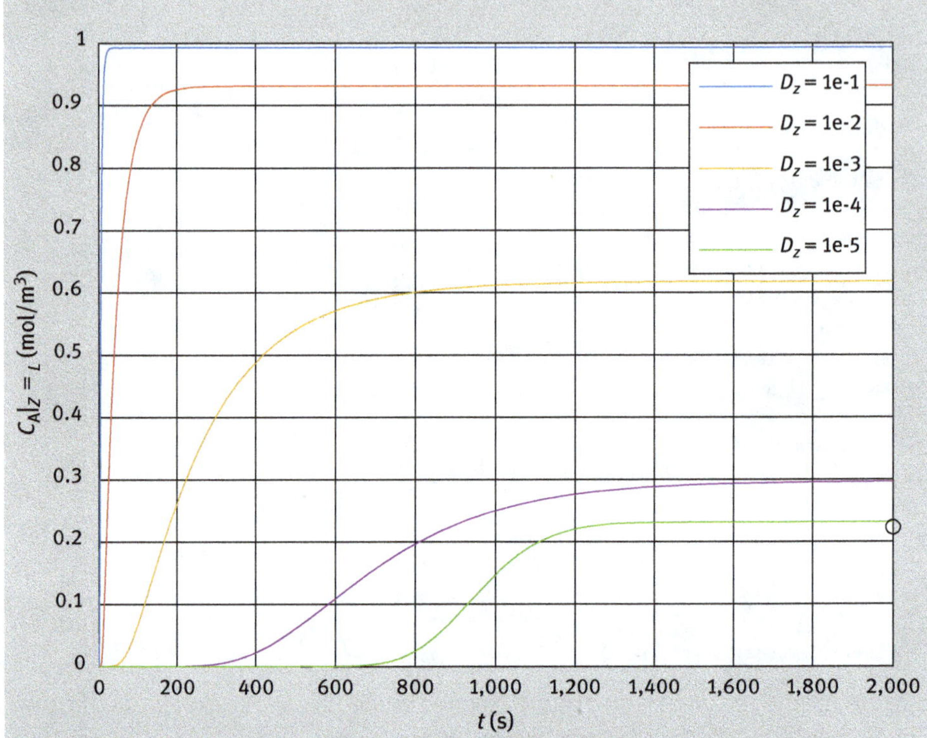

Figure 6.7: Concentration of the reactant at the outlet of the reactor as a function of time for different axial dispersion coefficients. The symbol represents the steady-state result obtained with the ideal PBR model.

As revealed, by decreasing the axial dispersion coefficient, the solution gets closer to the ideal PBR case, thus ranging from an ideal CSTR to PBR.

Radial dispersion is surely a key aspect to be considered when simulating larger-scale reactors, where radial gradients of both concentration and temperature are expected. In this case, the complexity of the models surely increases, leading to the definition of radial dispersion terms in both mass and energy balance equations

$$\frac{\partial C_i}{\partial t} = -u\frac{\partial C_i}{\partial z} + D_z\frac{\partial^2 C_i}{\partial z^2} + D_r\left(\frac{\partial^2 C_i}{\partial r^2} + \frac{1}{r}\frac{\partial C_i}{\partial r}\right) + \sum_{k=1}^{N_r} \nu_{ik} r_k \rho_B \tag{6.12}$$

$$\left(\sum_{i=1}^{N_C} C_i c_{p,i}\right)\frac{\partial T}{\partial t} = -\left(\sum_{i=1}^{N_C} C_i c_{p,i}\right)u\frac{\partial T}{\partial z} + \lambda_z\frac{\partial^2 T}{\partial z^2} + \lambda_r\left(\frac{\partial^2 T}{\partial r^2} + \frac{1}{r}\frac{\partial T}{\partial r}\right) + \sum_{k=1}^{N_r}(-\Delta_r H_k) r_k \rho_B \tag{6.13}$$

As revealed, the new terms include the occurrence of radial dispersion proportional by its coefficient D_r [m^2/s], and radial heat conduction proportional to the radial conductivity of the bed λ_r [J/(s m K)].

Each variable, that is, C_i and T, are solved along different coordinates, namely: (i) time; (ii) axis of the pipe; (iii) radius of the pipe. It is understandable that the general energy balance equation should not include the heat exchange term with the jacket as it is only present at $r = R$, condition that corresponding to one of the boundaries.

An appropriate set of boundary conditions must be defined to solve the partial differential equation (PDE) system. In general, it is possible to fix:

- $z = 0$: feed concentrations and temperature
- $z = L$: axial derivatives of concentrations and temperature are zero (the Danckwerts' closed boundary condition)
- $r = 0$: symmetry condition given by radial derivatives of concentrations and temperature as zero
- $r = R$: concentration and temperature derivatives are zero and heat flux takes place at the jacket

---

A system of such mathematical complexity requires the use of the method of lines, thus the discretization of both the axial and radial coordinate. Some hints for the solution of such a system are given in Chapter 3 of this book.

6.4 Packed bed with fluid–solid mass transfer resistance

Modeling a PBR where the fluid–solid mass transfer resistance is considered common practice when simulating the behavior of a lab-scale PBR, overall when facing with high viscous liquids, high residence times and high temperatures. As already seen in Chapter 3, in the mentioned conditions the fluid–solid mass transfer coefficient gets a low value; thus, it can happen that the dominating regime is not kinetics. A dynamic external diffusion model can be selected to describe the behavior of the mentioned reactor [3, 4]. The model consists of the simultaneous solution of two balance equations: (i) liquid bulk balance; (ii) solid surface balance.

The mass balance written on a thin bulk layer is reported as follows:

$$\frac{\partial C_i}{\partial t} = -u\frac{\partial C_i}{\partial z} + D_z\frac{\partial^2 C_i}{\partial z^2} - k_s a_{sp}(C_i - C_{i,s}), \quad u = \frac{\dot{V}}{A\varepsilon_B} \tag{6.14}$$

Equations (6.6) and (6.7) can be considered valid boundary conditions also in this case.

The pseudo-steady-state condition is considered as follows,

$$k_s a_{sp}(C_i - C_{i,s}) = \sum_{k=1}^{N_r} \nu_{ik} r_k \tag{6.15}$$

The mass transfer coefficient k_s can be estimated using existing correlations. For example, the Couret's correlation states that [5]

$$Sh_p = 5.4Re_p^{1/3}Sc^{1/4} \tag{6.16}$$

Which is valid for $0.04 < Re_p < 30$ and $1,700 < Sc < 11,000$, where the dimensionless groups are defined as it follows:

$$Sh_p = \frac{k_s d_p}{D} \tag{6.17}$$

$$Re_p = \frac{u d_p}{\nu} \tag{6.18}$$

$$Sc = \frac{\nu}{D} \tag{6.19}$$

An equivalent treatise could be made for the energy balance equation. The presented model is of outermost importance as a high number of applications are reported in the literature. An example of its application is provided as follows.

Example 6.6 Fluid–solid mass transfer resistance model

For a given reaction A → B, solve the PBR model taking into consideration the eventual fluid–solid mass transfer resistance. Compare with ideal PBR case. The settings needed are listed in Table 6.6.

Table 6.6: Settings needed to solve the simplified intraparticle PBR model.

Parameter	Value	Units
u	1e-3	m/s
D_e	1e-11	m^2/s
C_{feed}	1	mol/m^3
T	340	K
L	1	m
$k_m a_{sp}$	[1e-4,1e-3,1e-2]	1/s
k_{ref}	1e-6	$(m^3/kg)(m^3/mol)(1/s)$
T_{ref}	340	K
E_a	120	kJ/mol
ρ_B	1,500	kg/m^3

Matlab code for the solution of this example and results are reported as follows Figure 6.8.

```
% example 6.6
clc,clear

%% Reactor Settings
uf = 1e-3 ;       % m/s
rhoB = 1500 ;     % kg/m3
Cfeed = 1 ;       % mol/m3
T = 340 ;         % K
L = 1 ;           % m
kmasp = [1e-4,1e-3,1e-2]; % 1/s
zmesh=linspace(0,L,100);
tspan=linspace(0,2000,1e3);

%% Kinetics Settings
kref=1e-6;        % m3/(kg s)
Tref=340;         % K
Ea=120000;        % J/mol
R=8.3144;         % J/(K mol)

%% odecall
for i=1:length(kmasp)
    kmaspi=kmasp(i);
    f4=@(z,ca)odefun6_06(z,ca,uf,kref,Ea,R,Tref,rhoB,T,kmaspi);
    [z,ca] = ode15s(f4,zmesh,Cfeed);
    cax{i}=ca;
end

f4=@(z,ca)odefun6_06_Ideal(z,ca,uf,kref,Ea,R,Tref,rhoB,T);
[z,ca_Ideal] = ode15s(f4,zmesh,Cfeed);

%% Figures
plot(zmesh,cax{1},zmesh,cax{2},zmesh,cax{3},zmesh,ca_Ideal)
xlabel('\it z \rm [m]')
ylabel('\it C_{A} \rm [mol/m^{3}]')
legend('\it k_{m}a_{sp}\rm=1e-4','\it k_{m}a_{sp}\rm=1e-3','\it k_{m}a_{sp}\rm=1e-
2','Ideal PBR')
grid on

function [dcadz] = odefun6_06(z,ca,uf,kref,Ea,R,Tref,rhoB,T,kmasp)
options=optimset('Display','off');
fs=@(cas)zerofun6_06(cas,ca,kref,Ea,R,Tref,rhoB,T,kmasp);
cas=fsolve(fs,ca,options);
dcadz=-kmasp*(ca-cas)/uf;

end
```

```
function [F] = zerofun6_06(cas,ca,kref,Ea,R,Tref,rhoB,T,kmasp)
r=kinetic6_06(cas,T,kref,Tref,Ea,R); % mol/(kg s)

F=+rhoB*(-r)+kmasp*(ca-cas);
end
function [ reaz ] = kinetic6_06(ca,T,kref,Tref,Ea,R)
k=kref*exp(-Ea/R*(1/T-1/Tref));
reaz=k.*ca;

end

function [dcadz] = odefun6_06(z,ca,uf,kref,Ea,R,Tref,rhoB,T)
r=kinetic6_06(ca,T,kref,Tref,Ea,R); % mol/(kg s)

dcadz=rhoB*(-r)/uf;
end
```

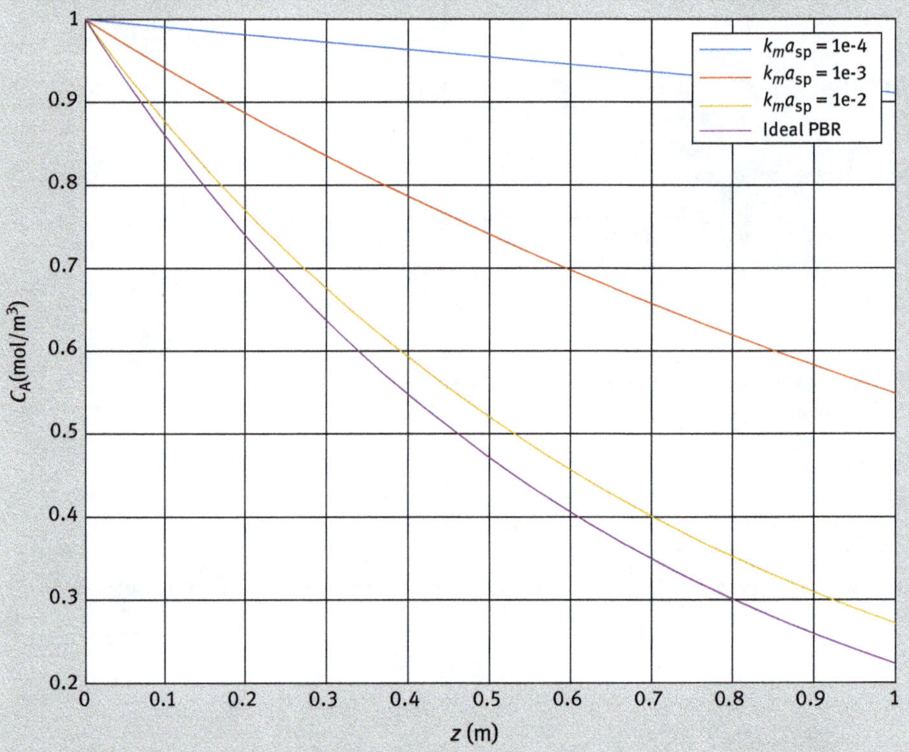

Figure 6.8: Concentration profiles as a function of the axial coordinate of the reactor for differ values of the fluid–solid mass transfer coefficient.

The overall rate is limited by low value of the mass transfer coefficient, and it approaches the ideality for high values.

6.5 Intraparticle diffusion model

If the intraparticle mass transfer is sufficiently limiting to rise concentration and temperature gradients, for example, high particle size or high viscosity of the fluid, the model must include the solution of the mass and energy balance equations in the catalyst particle. Writing a transient PBR model consists in defining a set of partial differential equations, given by the mass and energy balances for each phase [4]. The mass balance for one component present in the fluid phase is given as follows:

$$\frac{\partial C_i}{\partial t} = -u\frac{\partial C_i}{\partial z} + D_z\frac{\partial^2 C_i}{\partial z^2} + D_r\left(\frac{\partial^2 C_i}{\partial r^2} + \frac{1}{r}\frac{\partial C_i}{\partial r}\right) - N_{s,i} \tag{6.20}$$

The model is dynamic and both axial and radial dispersion effects are included. The net flux of a component i from the fluid to the solid phase must be considered

$$N_{s,i} = \frac{D_{e,i}s}{R_p}\frac{\partial C_{i,s}}{\partial r_p}\Big|_{r_p = R_p} \tag{6.21}$$

The mass balance for an arbitrary component i in the solid phase is defined in eq. (6.22):

$$\frac{\partial C_{i,s}}{\partial t} = \frac{D_{e,i}}{\varepsilon_p}\left(\frac{\partial^2 C_{i,s}}{\partial {r_p}^2} + \frac{s}{r_p}\frac{\partial C_{i,s}}{\partial r_p}\right) + \rho_p\frac{\varepsilon_f}{\varepsilon_p}\sum(v_{ik}r_k) \tag{6.22}$$

The geometry of a particle can be considered by adjusting the shape factor value (s).

The PDE system is hard to be solved, as one additional coordinate was added, that is, the particle radius. Central difference approximations could be used.

The set of boundary conditions is defined as follows:

- $z = 0$: feed concentrations and temperature
- $z = L$: axial derivatives of concentrations and temperature are zero (the Danckwerts' closed boundary condition)
- $r = 0$: symmetry condition given by radial derivatives of concentrations and temperature as zero
- $r = R$: concentration and temperature derivatives are zero and heat flux takes place at the jacket
- $r_p = 0$: symmetry given by zero derivatives of both concentration and temperatures
- $r_p = R_p$: continuity equations being the contact between the gas and the solid. No eventual film resistance was considered.

Several physical properties are needed for the computations. The pressure drop for gases can be calculated using the Ergun equation,

$$\Delta p/L = f_p \frac{\rho u^2}{d_\mathrm{p}} \left(\frac{1-\varepsilon_\mathrm{f}}{\varepsilon_\mathrm{f}{}^3} \right) \tag{6.23}$$

where the friction factor (f_p) is given by eq. (6.24):

$$f_p = \frac{150}{\mathrm{Re_p}} + 1.75, \ \mathrm{Re_p} = \frac{\rho u d_p}{(1-\varepsilon_\mathrm{f})\mu} \tag{6.24}$$

A simplified approach would be solving mass and energy balance equations, weighting the reaction rate term by the effectiveness factor:

$$u \frac{\partial C_i}{\partial z} = + \sum_{k=1}^{Nr} v_{ik} r_k \eta \tag{6.25}$$

Please check Chapter 3 for the effectiveness factor definition.

The effectiveness factor can be calculated solving the intraparticle mass balance equation with Method Of Line (MOL) approach, discretizing the particle coordinate.

First and second derivatives along the particle radius coordinate can be computed as it follows, using the centered finite difference method:

$$\frac{\partial C_{i,\mathrm{s}}}{\partial r_\mathrm{p}} \cong \frac{C_{i,\mathrm{s}}(j+1) - C_{i,\mathrm{s}}(j-1)}{2\Delta r} \tag{6.26}$$

$$\frac{\partial^2 C_{i,\mathrm{s}}}{\partial r_\mathrm{p}^2} \cong \frac{C_{i,\mathrm{s}}(j+1) - 2C_{i,\mathrm{s}}(j) - C_{i,\mathrm{s}}(j-1)}{\Delta r^2} \tag{6.27}$$

The application of the mentioned approach is reported in Example 6.7.

Example 6.7 PBR with intraparticle diffusion limitation

For a given reaction A → B, solve the intraparticle mass balance for a second order reaction kinetics, adopting the simplified approach for a PBR. The settings needed are listed in Table 6.7.

Table 6.7: Settings needed to solve the simplified intraparticle PBR model.

Parameter	Value	Units
u	1e-4	m/s
D_e	1e-11	m^2/s
C_feed	1e3	mol/m^3
T	340	K

Table 6.7 (continued)

Parameter	Value	Units
R_p	1e-2	m
L	1	m
k_{ref}	1e-9	$(m^3/kg)(m^3/mol)(1/s)$
T_{ref}	340	K
E_a	120	kJ/mol
ρ_B	1,500	kg/m^3

Matlab code for the solution of this example and results are reported as follows Figure 6.9:

```
% example 6.7
clc,clear

%% Reactor Settings
u = 1e-4 ;      % m/s

rhoB = 1500 ;     % kg/m3
Cfeed = 1e3 ;     % mol/m3
T = 340 ;         % K
L = 1 ;           % m

%% Kinetics Settings
kref=1e-9;      % (m3/kg)(m3/mol)(1/s)
Tref=340;       % K
Ea=120000;      % J/mol
R=8.3144;       % J/(K mol)

%% Particle Settings
De=1e-11; % m2/s
Rp=1e-2; % m

%% ode call
NN=100; % Number of inner points
zspan=linspace(0,L,1e2);
f1=@(z,V)odefun6_07(z,V,kref,Ea,R,Tref,u,rhoB,De,Rp,T,NN);
[z,V]=ode15s(f1,zspan,Cfeed);
```

```
ca=V(:,1);

for i=1:length(ca)
    cai=ca(i);
    [eta,cx]=eta_iso6_07(cai,De,Rp,kref,NN,T,Tref,Ea,R);
    etaz(i)=eta;
end
f1=@(z,V)odefun6_07_ideal(z,V,kref,Ea,R,Tref,u,rhoB,De,Rp,T,NN);
[z,V]=ode15s(f1,zspan,Cfeed);
ca_Ideal=V(:,1);

%% Figures
subplot(1,2,1)
plot(zspan,ca,'-r',zspan,ca_Ideal,'-b')
xlabel('\it z \rm [m]')
ylabel('\it C_{A} \rm [mol/m^{3}]')
legend('Intraparticle diffusion limitation','Ideal case')
axis([0 1 0 1000])
grid on

subplot(1,2,2)
plot(zspan,etaz,'-r')
xlabel('\it z \rm [m]')
ylabel('\it \eta \rm [-]')
grid on

function [dVdz] = odefun6_07(z,V,kref,Ea,R,Tref,u,rhoB,De,Rp,T,NN)
ca=V(1);
r=kinet_iso6_07(ca,T,kref,Tref,Ea,R); % mol/(kg s)
[eta,cx]=eta_iso6_07(ca,De,Rp,kref,NN,T,Tref,Ea,R);

%% Bulk phase
dVdz(1)=-r*rhoB*eta./u;

dVdz=dVdz';
end
function [dVdz] = odefun6_07_ideal(z,V,kref,Ea,R,Tref,u,rhoB,De,Rp,T,NN)
ca=V(1);
r=kinet_iso6_07(ca,T,kref,Tref,Ea,R); % mol/(kg s)
```

```
%% Bulk phase
dVdz(1)=-r*rhoB/u;
dVdz=dVdz';
end

function [ eta, cx ] = eta_iso6_07(cs,De,Rp,kref,NN,T,Tref,Ea,R)
%% discretization
dr=Rp/NN;
rx=0:dr:Rp;
ca0=0:cs*dr/Rp:cs-cs*dr/Rp; % initial assumption for ca profile
c0=[ca0'];

%% calculation of concentration profiles
options=optimoptions(@fsolve,'Display','off','TolFun',1e-20,...
        'PrecondBandWidth',inf);
fp=@(c)pellet_iso6_07(c,cs,De,dr,kref,NN,T,Tref,Ea,R);
[c, fc]=fsolve(fp,c0,options);

cax=[c(1:NN) ; cs(1)];
cx=[cax];

%% effectiveness factor
rsurf=kinet_iso6_07(cs,T,kref,Tref,Ea,R);
rr=kinet_iso6_07(cx,T,kref,Tref,Ea,R);
vp=(4/3)*pi*Rp^3;
fi=rr.*4.*pi.*rx'.^2;
eta=(1/vp)*trapz(rx,fi)/rsurf;
end

function [ fx ] = pellet_iso6_07(cx,cs,De,dr,kref,NN,T,Tref,Ea,R)
fx=zeros(NN,1);
%% reaction rates at discretization points
cax=[cx(1:NN); cs(1)];
c=cax;

%% reactor rate at all nodes
reaz=kinet_iso6_07(c,T,kref,Tref,Ea,R);

% balance on nodes
```

```
% node 1 - r=0
fx(1)=(cax(2)-cax(1))/dr;
% nodes j=2-(NN)
for j=2:NN
    der1=(cax(j+1)-cax(j-1))/(2*dr);
    der2=(cax(j+1)+cax(j-1)-2*    cax(j))/(dr^2);
    fx(j)=De*der2 + (2*De)/((j-1)*dr)*der1 - reaz(j);
end
end
function [ reaz ] = kinet_iso6_07(ca,T,kref,Tref,Ea,R)
k=kref*exp(-Ea/R*(1/T-1/Tref));
reaz=k.*ca.^2;
end
```

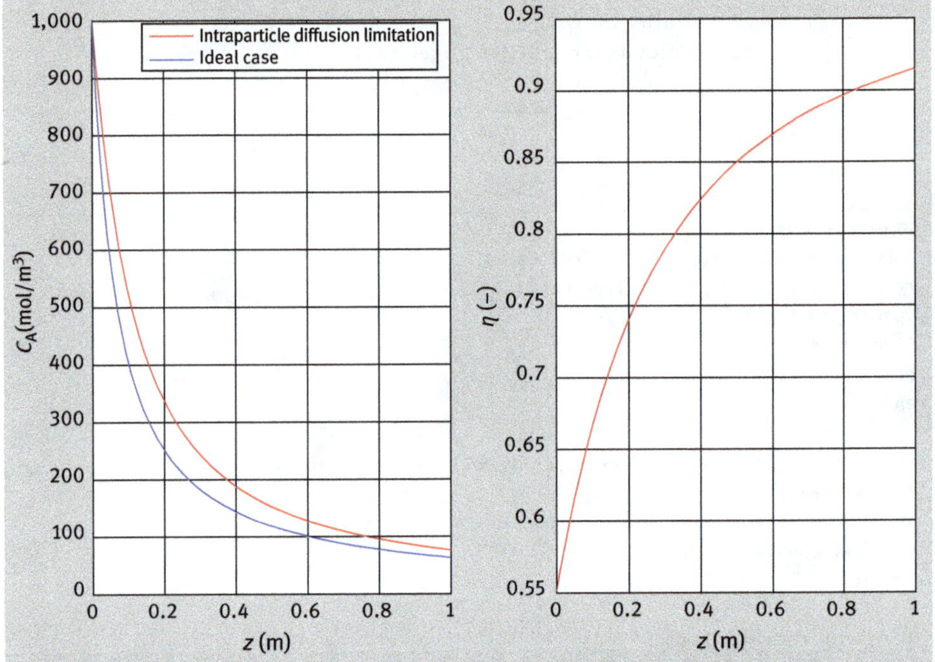

Figure 6.9: Concentration profiles as a function of the axial coordinate of the reactor (left). Effectiveness factor plot versus the reactor length (right).

As revealed, as the concentration profiles decreases, effectiveness factor increases approaching the unity. This is since the surface and internal reaction rates become very similar till reaching both zero when the reactant is totally consumed by the extent of the reaction. Moreover, the overall rate of the system is lower than the case where the intraparticle mass transfer limitation is considered, as the concentration profile of the reactant is characterized by a lower slope.

List of symbols

A	Catalyst surface	[m^2]
a_{sp}	Catalyst specific surface per fluid bulk volume	[m^2/m^3]
C_i	Concentration of component *i* in the bulk phase	[mol/m^3]
$C_{i,s}$	Concentration of component *i* in the solid phase	[mol/m^3]
$c_{p,i}$	Molar specific heat of component *i*	[J/(mol K)]
d_p	Particle diameter	[m]
D_i	Molecular diffusivity	[m^2/s]
$D_{e,i}$	Effective diffusivity	[m^2/s]
D_r	Radial dispersion coefficient	[m^2/s]
D_z	Axial dispersion coefficient	[m^2/s]
f_p	Friction factor	[–]
h	Fluid–solid heat transfer coefficient	[$W/(m^2 K)$]
k_m	Fluid–solid mass-transfer coefficient of component *i*	[m/s]
L	Bed length	[m]
N_c	Number of components	[–]
N_r	Number of reactions	[–]
p	Pressure	[Pa]
r	PBR radial coordinate	[m]
r_k	Reaction rate of reaction *k*	[mol/(kg s)]
r_p	Particle radial coordinate	[m]
Re_p	Reynolds particle number	[–]
s	Shape factor	[–]
Sc	Schmidt number	[–]
Sh	Sherwood particle number	[–]
t	Time	[s]
T	Temperature	[K]
T_J	Jacket temperature	[K]
u	Flow interstitial velocity	[m/s]
U	Overall heat transfer coefficient	[J/(m s K)]
V	Bed volume	[m^3]
$\dot{V}$	Fluid flowrate	[m^3/s]
w_{cat}	Catalyst mass	[kg]
z	PBR length coordinate	[m]

Greek letters

$\Delta_r H_k$	Reaction enthalpy	[J/mol]
ε_B	Bed porosity	[–]
ε_f	Fluid void fraction	[–]
ε_p	Particle porosity	[–]
λ_r	Radial heat conductivity	[J/(m s K)]
λ_z	Axial heat conductivity	[J/(m s K)]
ρ	Fluid density	[kg/m^3]
ρ_B	Catalyst bulk density	[kg/m^3]

ρ_p	Density of the catalyst	[kg/m^3]
$v_{i,k}$	Stoichiometric coefficient of component *i* in reaction *k*	[–]
v	Fluid viscosity	[m^2/s]
μ	Fluid viscosity	[Pa s]

References

[1] T. Kilpiö, E. Behravesh, V. Russo, K. Eränen, T. Salmi. Physical modelling of the laboratory-scale packed bed reactor for partial gas-phase oxidation of alcohol using gold nanoparticles as the heterogeneous catalyst. Chemical Engineering Research and Design 2017, 117, 448–459.

[2] T. Kilpiö, V. Russo, K. Eränen, T. Salmi. Design and modeling of laboratory scale three phase fixed bed reactors. Physical Sciences Reviews 2016, 1(3), Online open access journal. doi:10.1515/psr-2015-0020.

[3] A.P. Nebreda, V. Russo, M. Di Serio, K. Eränen, D.Y. Murzin, T. Salmi, H. Grénman. High purity fructose from inulin with heterogeneous catalysis – from batch to continuous operation. Journal of Chemical Technology & Biotechnology 2019, 94, 418–425.

[4] V. Russo, T. Kilpiö, M. Di Serio, R. Tesser, E. Santacesaria, D.Y. Murzin, T. Salmi. Dynamic non-isothermal trickle bed reactor with both internal diffusion and heat conduction: Arabinose hydrogenation as a case study. Chemical Engineering Research and Design 2015, 102, 171–185.

[5] F. Coeuret. L'electrode poreuse percolante (epp)-I. Transfert de matiere en lit fixe. Electrochimica Acta 1976, 21(3), 185–193.

Chapter 7
Parameters estimation

7.1 Introduction

The general task of parameters estimation is encountered very frequently when laboratory or plant experimental data must be used to develop and validate a physicomathematical model to be employed for scale-up purposes [1, 2]. A model that adequately represents the system characteristics, contains some parameters that cannot be calculated a priori by means of suitable correlations or found in the literature and then must be evaluated on the basis of experimental observations. The process of using the observations derived from experimental plan to estimate unknown parameters in the developed mathematical models is usually performed on statistical basis as the number of parameters is usually much less than experimental data. The model under development consists of a group of equations (algebraic or differential) in which a finite set of unknown parameters appears. Fundamentally, the parameters estimation approach could be based on two different strategies or algorithms: (i) deterministic algorithms, in which the minimization of the objective function is mainly based on gradient of the objective function itself; (ii) stochastic or evolutive algorithms (EAs), where the minimization of the objective function is pursued through a complex evolutive strategy in which parameters are gradually updated on stochastic base.

In the field of parameter estimation, a particular attention must be payed to the use of an objective function which global minimum must be searched for. Usually, the objective function is built as root mean square (RMS) quadratic error between the model response and experimental system response. In this case the principle of least squares, or weighed least squares, is applied and the error function (objective function) is the sum of squares of the differences between experiments and model response (residuals).

7.2 Common minimization strategies

As we have briefly discussed in the introduction section, two main classes of algorithms are essentially available and widely adopted in the literature for solving the problem of searching for the minimum of an objective function in a least squares linear or nonlinear regression.

In the first class, that of deterministic algorithms, comprised the following [3–7]:
- gradient-based methods
- simplex
- trust-region algorithm
- generalized reduction of gradient

https://doi.org/10.1515/9783110632927-007

The second class is related to stochastic and evolutionary algorithms (EAs) like the following [8–10]:
- particle swarm optimization (PSO)
- genetic algorithm optimization
- pattern search optimization
- ant colony optimization

In Matlab developing environment, practically all these algorithms are present, also as a combination of two or more of them. To the user, generally, is left the task of writing only the driver code and the function that implement the objective to be minimized.

7.2.1 Deterministic minimization algorithms

Deterministic minimization algorithms were developed and applied, from the very beginning, to the problems of function minimization and nonlinear parameter estimation. The majority of them is based on function gradient evaluation and then can be applied to continuous and differentiable objective function. On the contrary, some of these algorithms are not based on gradient, for example, the simplex algorithm [11–13], and can be more broadly employed to solve minimization problems.

As an example, we report here the bases for a generic gradient minimization method. The general form of a mathematical model can be represented as in the following equation:

$$\hat{y} = f(x, \beta) \tag{7.1}$$

where f is the function defining the model (algebraic or differential); x is the vector of the Ns experimental observations, independent variables; β the vector of p unknown parameters that must be estimated; $\hat{y}$ the vector of calculated dependent variables (response of the model); y_i the vector of experimental dependent variables; $\bar{y}$ the average value of experimental y data.

In a nonweighed least squares approach, the usual objective function can be written as residuals sum of squares:

$$O(\beta) = \sum_{i=1}^{Ns} (y_i - \hat{y}_i)^2 \tag{7.2}$$

Or, alternatively, in the case of a weighed least squares:

$$O(\beta) = \sum_{i=1}^{Ns} w_i (y_i - \hat{y}_i)^2 \tag{7.3}$$

Where w_i is the weight attributed to the *i*th experimental measure. In order to find the minimum of the objective function, the function itself must be differentiable with respect to the unknown parameters and a system of partial derivatives can be written as follows:

$$\frac{\partial O(\beta)}{\partial \beta_j} = \frac{\partial}{\partial \beta_j}\left[\sum_{i=1}^{Ns}(y_i - \hat{y}_i)^2\right] = 0 \; j = 1 \ldots p \qquad (7.4)$$

If the mathematical model contains *p* adjustable parameters, the system earlier is represented by *p* partial derivatives that must be simultaneously solved. The solution of this type of system can be accomplished, for example, with algorithms like Newton–Raphson for simultaneous algebraic equations.

7.2.2 Evolutive minimization algorithms

EAs are a broad class of stochastic optimization algorithms inspired by biological processes and, more in particular, by those processes that allows populations of organisms to develop a strategy for adapting to the surrounding environment. With the reference to the genetic species evolution, this concept is focused in the survival of the fittest. EAs can be considered a special type of artificial intelligence and are motivated by optimization processes that we observe in nature, such as natural selection, species migration, bird swarms, human culture, and ant colonies.

EAs-based methods of global minimum search have some advantages over the other gradient-base methods and, in particular, the two most important ones are as follows [3, 4]:

- They can be applied to optimization problems (minimization or maximization) that consists of discontinuous, nondifferentiable and nonconvex objective function and constraints.
- They can rather easily avoid falling, during the iterative search, into local optima and are then more efficient in finding global minimum.

On the other hand, the main drawback of EAs is that with these algorithms no information can be obtained regarding statistics and confidence intervals (see Section 7.4) as they are not based on gradient evaluation. However, in the optimization practice, the user can decide to exploit the ability of EAs to obtain the global minimum and, once the minimum is located, only few iterations with gradient-based method can be further performed gaining also statistical information.

This type of minimization strategy was developed relatively recently but the detailed description of all these algorithms is outside the scope of the present book and only a brief description and concepts of PSO are presented here as explicative example.

The original PSO algorithm [14] was inspired, as many EAs, by the behavior of biological organisms with the ability to work as a group in finding the best resources (minimum or maximum) in a specified area. The seeking strategy of a group of such organisms can be exploited in the optimization search for a solution of nonlinear model in the real space of multiple parameters. During their search for the minimum, each particle of the swarm can communicate with the others, sharing in this way the obtained result in terms of objective function minimization. Each individual particle is connected to all the other particle in the swarm and can obtain useful information from the other. To each single ith particle, three vectors are associated as described in the following:

- Particle position: a vector in p-dimensional search space (the dimension of this space is the number of adjustable parameters):

$$\vec{x}_i = (x_{i1}, x_{i2}, x_{i3}, \ldots, x_{ip}) \tag{7.5}$$

- The vector of the best position that the ith particle have found in his search:

$$\vec{p}_i = (p_{i1}, p_{i2}, p_{i3}, \ldots, p_{ip}) \tag{7.6}$$

- The particle velocity vector:

$$\vec{v}_i = (v_{i1}, v_{i2}, v_{i3}, \ldots, v_{ip}) \tag{7.7}$$

At the beginning of the search procedure, both particle position and particle velocity are initialized randomly. In particular, particles are distributed randomly in the search space defined between the lower and the upper bound of the adjustable parameters. The iterative search for the minimum proceeds then by moving the particle swarm in the search space by simply updating the particle position and velocities through the following recursive relations:

$$\begin{aligned} v_{i,k} &= v_{i,k} + Ce_1(p_{i,k} - x_{i,k}) + Ce_2(p_{g,k} - x_{i,k}) \quad k = 1 \ldots p \\ x_{i,k} &= x_{i,k} + v_{i,k} \quad k = 1 \ldots p \end{aligned} \tag{7.8}$$

where C is a constant (2.0 in the original algorithm formulation), e_1 and e_2 are two random numbers independently generated at each update and $\vec{p}_g$ is the best global result obtained by the entire swarm of the neighbor of the particle. Originally, the particles velocities were limited to a maximum value to avoid "swarm explosion" that in some case could occurs.

7.3 Objective function definition

As we have seen in Section 7.2, the minimization algorithms, deterministic or evolutive, have the scope to locate the global minimum of a suitable objective function that depends on a set of p adjustable parameters. This objective function is usually defined as in the eqs. (7.2) or (7.3) for, respectively, unweighed or weighed least squares problems in which the residuals sum of squares must be minimized. As alternatives to this choice, the following objective functions could be adopted, representing for example the average square error or the average relative error:

$$O(\beta) = \frac{1}{N_S}\sum_{i=1}^{Ns}(y_i - \hat{y}_i)^2 \tag{7.9}$$

$$O(\beta) = \frac{1}{N_S}\sum_{i=1}^{Ns}\left(\frac{y_i - \hat{y}_i}{y_i}\right)^2 \tag{7.10}$$

In terms of minimization algorithm all these choices are practically equivalent. More care must be adopted when the minimization problem involves the simultaneous fitting of different variables, for example temperature and conversion in a kinetic study. In this case a two-contributions objective function can be suitable as in the following expression:

$$O(\beta) = \frac{1}{N_S}\left[w_x\sum_{i=1}^{Ns}\left(\frac{x_i - \hat{x}_i}{x_i}\right)^2 + w_T\sum_{i=1}^{Ns}\left(\frac{T_i - \hat{T}_i}{T_i}\right)^2\right] \tag{7.11}$$

where x and T are, respectively, the conversion and the temperature experimental and calculated; w_x and w_T are two weights that must be defined in order to make comparable the value of the two part of the objective function (scaling problem). Without the definition of these two scaling parameters, the minimization algorithm is addressed to optimize the temperature data in a prevalent way as the residual between experimental and calculated temperature have much higher values.

7.4 Statistics: confidence intervals, covariance and correlation matrix

For a specific algebraic or differential model, after the estimation of the adjustable parameters, other important questions could arise: how much reliable the estimated parameters are? What is the error or confidence interval on them? Are these parameters correlated among them (i.e., the value of one parameter influences the value of the other)?

The answers to these fundamental questions represent a statistical issue and can be faced by adopting a standardized mathematical procedure [15, 16]. As the estimation approach is based frequently on least square minimization, we refer to a generic model already defined as in eq. (7.1). The problem is then to search for a set β of p unknown adjustable parameters that minimize a suitable objective function such as.

$$\min \sum_{i=1}^{n} [y_i - f(x, \beta)]^2 \tag{7.12}$$

In expression (7.12) the calculated value of the model response, corresponding to the ith experimental y data, is a function of the whole x vector and, also, of the whole set of adjustable parameters. The first step for the evaluation of statistics information about adjustable parameters consists of the calculation of the parameter's standard deviation. The evaluation of the standard deviation of the parameters estimated values can be calculated with a mathematical procedure that starts from the Jacobian matrix calculated as follows:

$$J_{i,j} = \frac{\partial f(x_i, \beta)}{\partial \beta_j} \, i = 1, \ldots, n \; j = 1, \ldots, p \tag{7.13}$$

The resulting J matrix is made of n rows (number of experimental observations) and p columns (number of parameters). From this Jacobian matrix, the covariance matrix can be directly evaluated, through standard matrix operations, with the following relation:

$$V = s^2 \left[J^T J\right]^{-1} \tag{7.14}$$

Where s is the RMS error, named also standard deviation of the residues, defined as follows:

$$s = \sqrt{\frac{\sum_{i=1}^{n} (y_i - \hat{y}_i)^2}{n - p}} \tag{7.15}$$

The covariance matrix V, defined by eq. (7.15), is a fundamental statistical quantity that can be employed directly in the estimation of the standard error on each parameter of the model as the square root of diagonal elements of V:

$$s_j = \sqrt{V_{jj}} \; j = 1, \ldots, p \tag{7.16}$$

In many cases, instead of parameters standard deviation s_j, we are more interested in confidence intervals on parameter estimated values. These intervals represent much better the accuracy or reliability of parameters and are usually symmetric and centerd on the estimated value of each parameter and with an amplitude that is proportional to the chosen probability. The concept of confidence interval is indeed a statistical concept and represents the interval, around the estimated value of the parameter, inside which the true value of the parameter can be found with a certain probability. The more the probability level is high (closer to 1), and consequently the lower is the risk of missing the true value, the wider is the corresponding confidence interval for that specific parameter. Usually, a probability level of 95% is chosen (corresponding to a risk of $\alpha = 5\%$) and this means that the confidence interval can be evaluated as a product of the standard error on the parameter (s_j eqs. (7.6)) and a multiplying factor evaluated on statistical basis: usually the inverse Student's t-distribution is used for the calculation of this last multiplier factor as follows:

$$t_{\text{inv}}\left(1-\frac{\alpha}{2},n-p\right) \tag{7.17}$$

The quantity n–p represents the degrees of freedom of the system evaluated as the difference between the number of experiments (n) and the number of the parameters in the model (p). The complete expression for associating a confidence interval to each adjustable parameter is therefore:

$$\beta_j \pm s_j \cdot t_{\text{inv}}\left(1-\frac{\alpha}{2},n-p\right) \tag{7.18}$$

From this relation, it is evident that the lower is the chosen risk (the higher is the probability), the wider is the confidence interval on the parameters of the model as Student's t-function increase in the same sense.

In some cases, high values of confidence intervals can be obtained involving an error band on the two sides of the corresponding parameter that cross the zero. If the parameters are intrinsically positive, the value of zero must be chosen as lower limit giving place to an asymmetric error band.

When the mathematical model becomes more and more complex, and when the number of adjustable parameters is relatively high, another useful information is available from statistical considerations. This tool is the correlation matrix that could give us quantitative information about the correlation between parameters. When some parameters are correlated among them, the possibility to reformulate the model should be considered in order to eliminate the correlation itself. Each element in the correlation matrix can be evaluated from the corresponding elements of the covariance matrix V, according to the following equation:

$$\text{Corr}_{ij} = \frac{V_{ij}}{\sqrt{V_{ii}}\sqrt{V_{jj}}} \tag{7.19}$$

The correlation matrix is characterized by diagonal elements equal to 1 while each off-diagonal element (i,j) represents the correlation between the corresponding couple of parameters i and j. The more an off-diagonal element is close to 1, the highest is the correlation between the corresponding parameters i and j. Obviously, the correlation matrix is symmetrical with respect to the diagonal.

Example 7.1 Single reaction, batch reactor, determination of single kinetic constant plot with confidence band on prediction
In a batch reactor a single liquid-phase chemical reaction occurs at a constant temperature, according to the following simple scheme:

$$A \rightarrow B$$

The kinetic of this reaction was studied by collecting experimental concentration data for A and B along the reaction time. The collected experimental data are summarized in Table 7.1.

Table 7.1: Experimental data concentration versus time, Example 7.1.

Time (s)	Ca (mol/m³)	Cb (mol/m³)
0	15.00	0.00
100	12.51	3.15
200	9.05	6.55
400	5.11	9.83
600	3.00	12.05
800	1.29	13.51
1,000	0.68	14.08

The reaction rate, for the considered reaction, is expressed by a first-order expression:

$$r = kC_A$$

The differential mass balance related to A and B can be written then in the usual form:

$$\frac{dC_A}{dt} = -r$$

$$\frac{dC_B}{dt} = +r$$

Develop a Matlab code that allows the determination of the unknown kinetic constant k with the associated confidence interval. Draw a plot in which experimental and calculated values are reported together with the confidence bands on the predicted values of Ca and Cb.

Matlab code for the solution of this example and results are reported as follows Figure 7.1:

```
% example 7.1
clc,clear
data=[ 0 15.00 0.00;
    100 12.51 3.15;
    200 9.05 6.55;
    400 5.11 9.83;
    600 3.00 12.05;
    800 1.29 13.51;
    1000 0.68 14.08];
```

```
ts = data(:,1);
CAs = data(:,2);
CBs = data(:,3);
C0=[15 0]; % initial concentrations

%% parameter estimation
objf=@(kp)obj_ex8_01(kp,data,C0);
k0 = [0.01]; % initial kinetic constants
lb = [ 0 ];
ub = [100];
options=optimoptions(@lsqnonlin,'Display','Iter');
[k,resnorm,residual,exitflag,output,lambda,jacobian]= . . .
                    lsqnonlin(objf,k0,lb,ub,options);
ci = nlparci(k,residual,'jacobian',jacobian);
CIk=abs(ci(:,1)-ci(:,2))/2;
Jacobian=full(jacobian);
s2=resnorm/length(ts);
covarM=s2*inv(Jacobian'*Jacobian) ;
corrM=corrcov(covarM) ;
stdev=sqrt(diag(covarM));
disp('-------------------------------')
disp('------- Fitted Parameter --------')
disp('-------------------------------')
fprintf(' Parameter k %1.0f = %10.5e C.I.: %10.5e \n',1,k,CIk)
%% simulation
tspan=0:0.1:1200; % time range for integration (s)
ode=@(t,C)ode_ex8_01(t,C,k);
[tx,Cx]=ode15s(ode,tspan,C0);
modelf=@(k,tspan)model_ex8_01(k,tspan,data,C0);
[Cpred,delta]=nlpredci(modelf,tspan,k,residual,'Jacobian',jacobian);
L=length(Cpred);
CApred=Cpred(1:L/2);
CBpred=Cpred(L/2+1:end);
deltaA=delta(1:L/2);
deltaB=delta(L/2+1:end);
CAp = CApred+deltaA;
CAm = CApred-deltaA;
CBp = CBpred+deltaB;
CBm = CBpred-deltaB;

%% plot
plot(tx,Cx(:,1),'r-',tx,Cx(:,2),'b-', . . .
    ts,CAs,'ro',tspan,CAp,'r-',tspan,CAm,'r-', . . .
    ts,CBs,'bo',tspan,CBp,'b-',tspan,CBm,'b-')
grid
xlabel('Time (s)')
ylabel('Concentraitons (mol/m3)')
legend('A','B')
```

```
function [res] = obj_ex8_01(kp,data,C0)
ts = data(:,1);
tspan = ts;     % time range for integration (s)
ode=@(t,C)ode_ex8_01(t,C,kp);
[tc,Cc]=ode15s(ode,tspan,C0);
resA = data(:,2)-Cc(:,1);
resB = data(:,3)-Cc(:,2);
res=[resA;resB];
end

function [Ch] = model_ex8_01(kp,ts,data,C0)
tspan = ts;     % time range for integration (s)
ode=@(t,C)ode_ex8_01(t,C,kp);
[tc,Cc]=ode15s(ode,tspan,C0);
CAc=Cc(:,1);
CBc=Cc(:,2);
Ch=[CAc;CBc];
end

function [dC] = ode_ex8_01(t,C,k)
r=k*C(1);
dC(1)=-r;
dC(2)=+r;
dC=dC';
end
```

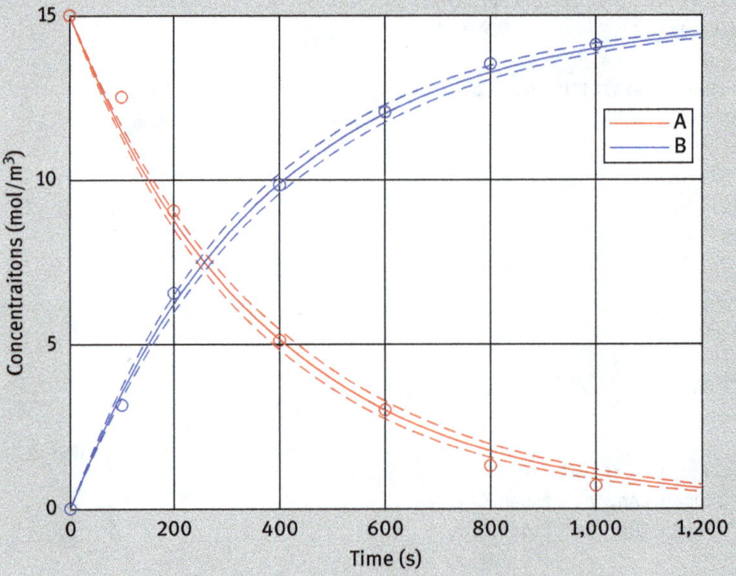

Figure 7.1: Experimental and calculated concentration of A and B versus time. Dots are experimental values, continuous curves represent model prediction and dashed curves are confidence bands.

```
                      Norm of   First-order
Iteration Func-count f(x)  step        optimality
    0       2     249.259                 140
    1       4     36.2657 0.0592468        76.7
    2       6     2.18576 0.0200866        8.88
    3       8     1.82333 0.00215113    8.33e+03
    4      10     1.82325 2.79097e-07     0.0148
-------------------------
------ Fitted Parameter ------
-------------------------
Parameter k 1=2.68213e-03 C.I.: 1.48363e-04
```

Example 7.2 Multiple reactions, liquid-phase batch reactor, parameter fitting, correlation and covariance matrix, sensitivity study

In a batch reactor a set of four multiple reactions, in a complex scheme, occur involving five components identified as A, B, C, D and E. The reactions scheme is the following:

$$A + B \longleftrightarrow 2C + D$$
$$A + C \longrightarrow E$$
$$D \longrightarrow 2B$$

The first reaction is reversible so the forward and reverse reactions must be accounted for as two separate reactions. On the basis of this scheme, the following stoichiometric matrix can be arranged in which the stoichiometric coefficients of all the components in all the reactions are included.

Table 7.2: Stoichiometric matrix for Example 7.2.

Component	Reaction 1	Reaction 2	Reaction 3	Reaction 4
A	−1	+1	−1	0
B	−1	+1	0	+2
C	2	−2	−1	0
D	1	−1	0	−1
E	0	0	1	0

The kinetics of this reactive system was studied in a batch reactor with a volume of 0.003 m^3 in isothermal conditions. The following experimental data, representing the moles of each components as function of time, have been collected:

Table 7.3: Experimental data mole number versus time, Example 7.2.

Time (s)	A (mol)	B (mol)	C (mol)	D (mol)	E (mol)
0	1.500	1.200	0.000	0.000	0.000
100	0.951	0.973	0.203	0.231	0.254
200	0.753	0.905	0.160	0.304	0.431
400	0.515	0.831	0.124	0.365	0.610
600	0.385	0.795	0.095	0.398	0.708
800	0.297	0.779	0.085	0.426	0.773
1,000	0.238	0.765	0.075	0.443	0.814

The expressions for the four reaction rates can be expressed as follows:

$$r_1 = k_1 C_A C_B$$

$$r_2 = k_2 C_C^2 C_D$$

$$r_3 = k_3 C_A$$

$$r_4 = k_4 C_D$$

In the previous expressions, four unknown kinetic constants are present that must be determined through fitting on the experimental data. The reactor was initially charged with 1.5 mol of component A and 1.2 mol of B. In a compact form, the material balance for batch reactor is as follows:

$$\frac{dn_i}{dt} = V_R \sum_{k=1}^{Nr} v_{i,k} r_k \quad i = \text{A, B, C, D, E} \quad N_R = \text{four reactions}$$

The purpose of the present solved example is to evaluate by fitting the four adjustable parameters (kinetic constants) and the related confidence intervals, covariance and correlation matrix. Evaluate also the sensitivity of the objective function with respect to each individual parameter.

Matlab code for the solution of this example and results are reported as follows Figure 7.2:

```
% example 7.2
clc,clear
data=[ 0 1.500 1.200 0.000 0.000 0.000;
    100 0.951 0.973 0.203 0.231 0.254;
    200 0.753 0.905 0.160 0.304 0.431;
    400 0.515 0.831 0.124 0.365 0.610;
    600 0.385 0.795 0.095 0.398 0.708;
    800 0.297 0.779 0.085 0.426 0.773;
    1000 0.238 0.765 0.075 0.443 0.814];

ts = data(:,1);
nas = data(:,2);
```

```
nbs = data(:,3);

ncs = data(:,4);
nds = data(:,5);
nes = data(:,6);

Nc=5;            % number of components (-)
Nr=4;            % number of reactions (-)
Vr=0.003;        % reaction volume (m3)
Np=4;            % number of parameters (-)
ni=[-1 +1 -1 0;  % matrix of stoichiometric coefficients
  -1 +1 0 +2; % row -> component
  +2 -2 -1 0; % column -> reaction
  +1 -1 0 -1
  0 0 +1 0];
n0=[1.5 1.2 0 0 0]; % initial moles (mol)

%% parameter estimation
objf=@(kp)obj_ex8_02(kp,data,n0,Vr,Nc,Nr,ni);
k0 = [0.01 0.02 0.3 0.01]/1000; % initial kinetic constants
lb = [ 0 0 0 0];
ub = [10 10 10 10];
options=optimoptions(@lsqnonlin,'Display','Iter');
[k,resnorm,residual,exitflag,output,lambda,jacobian]=lsqnonlin(objf,k0,lb,ub,
options);
ci = nlparci(k,residual,'jacobian',jacobian);
CIk=abs(ci(:,1)-ci(:,2))/2;
Jacobian=full(jacobian);
s2=resnorm/length(ts);
covarM=s2*inv(Jacobian'*Jacobian) ;
corrM=corrcov(covarM) ;
stdev=sqrt(diag(covarM));

disp('-------------------------')
disp('------ Fitted Parameters -------')
disp('-------------------------')
for j=1:Np
  fprintf(' Parameter k %1.0f = %10.5e C.I.: %10.5e \n',j,k(j),CIk(j))
end
disp(' ')
disp('-------------------------')
disp('------ Covariance Matrix -------')
disp('-------------------------')
disp(covarM)
disp('-------------------------')
disp('-------------------------')
disp('------ Correlation Matrix -------')
disp('-------------------------')
```

```
disp(corrM)
disp('------------------')
%% simulation
tspan=0:0.1:1200; % time range for integration (s)
ode=@(t,n)ode_ex8_02(t,n,k,Vr,Nc,Nr,ni);
[tx,nx]=ode15s(ode,tspan,n0);
%% sensitivity
kopt=k;
for jk=1:4
  ks=linspace(kopt(jk)*0.2,kopt(jk)*4);
  kks(:,jk)=ks;
  k_sens=kopt;
  for j=1:100
    k_sens(jk)=ks(j);
    res = obj_ex8_02(k_sens,data,n0,Vr,Nc,Nr,ni);
    Fo(j,jk)=sum(res.^2);
  end
end
kf=linspace(0.2,4);

%% plot
subplot(1,3,1)
plot(tx,nx(:,1),tx,nx(:,2),tx,nx(:,3), . . .
  ts,nas,'ro',ts,nbs,'ko',ts,ncs,'bo')
grid
xlabel('Time (s)')
ylabel('Moles (mol)')
legend('A','B','C')
subplot(1,3,2)
plot(tx,nx(:,4),tx,nx(:,5), . . .
    ts,nds,'go',ts,nes,'mo')
grid
xlabel('Time (s)')
ylabel('Moles (mol)')
legend('D','E')
subplot(1,3,3)
semilogy(kf,Fo)
grid
xlabel('Parameter factor (-)')
ylabel('Objective function')
legend('k1','k2','k3','k4')

function [res] = obj_ex8_02(kp,data,n0,Vr,Nc,Nr,ni)
ts = data(:,1);
tspan = ts; % time range for integration (s)
ode=@(t,n)ode_ex8_02(t,n,kp,Vr,Nc,Nr,ni);
[tc,nc]=ode15s(ode,tspan,n0);
ndat=length(tc);
```

```
nres=0;
for j=1:5
  for k=1:ndat
    nres=nres+1;
    res(nres)=data(k,j+1)-nc(k,j);
  end
end
end

function [dn] = ode_ex8_02(t,n,k,Vr,Nc,Nr,ni)
c=n/Vr;
r(1)=k(1)*c(1)*c(2);
r(2)=k(2)*c(3)^2*c(4);
r(3)=k(3)*c(1)*c(3);
r(4)=k(4)*c(4);
for j=1:Nc
  dn(j)=0;
  for jj=1:Nr
    dn(j)=dn(j)+Vr*r(jj)*ni(j,jj);
  end
end
dn=dn';
end

                                Norm of     First-order
Iteration Func-count  f(x)        step        optimality
0       5     0.747405                       8.63e+04
1       10    0.267199   2.22736e-05         1.37e+04
2       15    0.130385   3.42533e-05         2.14e+03
3       20     0.11241   0.00369201            93
4       25     0.11241   0.00729118            93
5       30    0.108903    0.0018228           112
6       35    0.100852   0.00364559        1.75e+03
7       40   0.0856373   0.00476319        1.07e+04
8       45   0.0636562   0.00243488        1.61e+03
9       50   0.0636562   0.00487993        1.61e+03
10      55   0.0544271   0.00121998          0.0342
11      60    0.036057   0.00243996        7.06e+03
12      65    0.036057   0.00467212        8.06e+03
13      70   0.0256018   0.00116803        1.07e+04
14      75   0.0233221   0.00233606        3.07e+04
15      80  0.00763289    1.4729e-05        2.26e+04
16      85  0.00223526   1.20085e-05           328
17      90  0.00221192   0.000230367           387
18      95  0.00219159   0.000584016           111
19      100  0.00217465   0.00048649           240
20      105  0.00217465   9.40892e-05          240
21      110  0.00217282   2.35223e-05          363
22      115  0.00217231   1.15472e-05        0.591
```

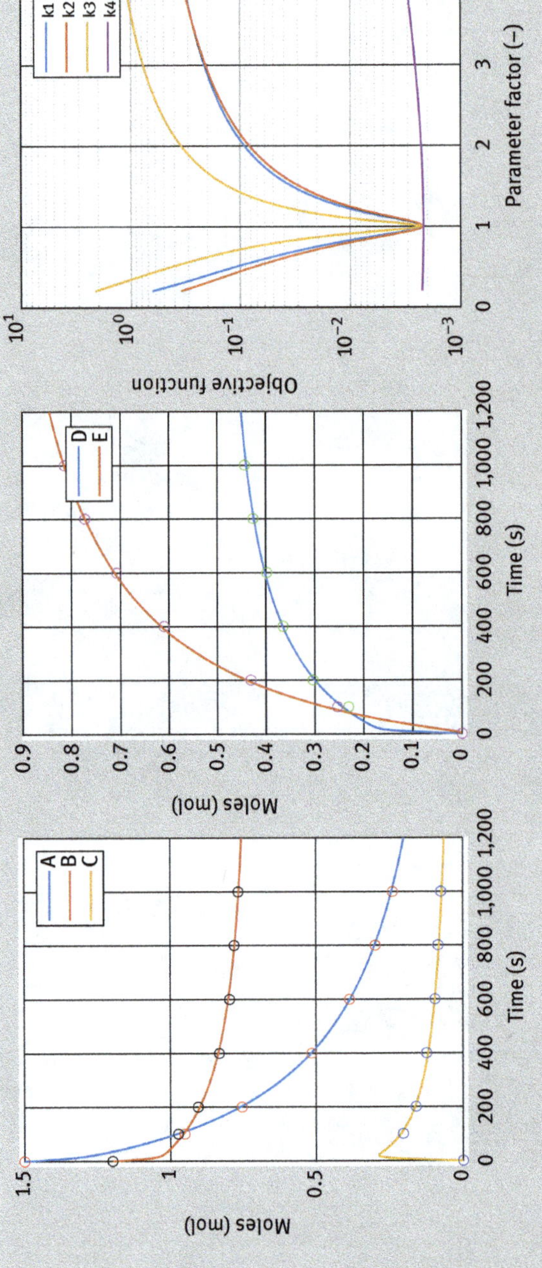

Figure 7.2: (Left) Experimental and calculated mole numbers for A, B and C versus time; (center) experimental and calculated mole numbers for D and E; (right) sensitivity study: effect of k1-k4 values on objective function. Dots are experimental values, continuous curves represent model prediction.

```
Local minimum possible.
lsqnonlin stopped because the final change in the sum of squares relative to
its initial value is less than the value of the function tolerance.

<stopping criteria details>
-------------------------------------------
------------ Fitted Parameters -------------
-------------------------------------------
Parameter k 1 = 2.67734e-05 C.I.: 1.89680e-05
Parameter k 2 = 6.36766e-06 C.I.: 4.89266e-06
Parameter k 3 = 3.05121e-05 C.I.: 1.51908e-06
Parameter k 4 = 9.59731e-06 C.I.: 1.67416e-05

-------------------------------------------
----------- Covariance Matrix --------------
-------------------------------------------
1.0e-09 *

 0.3830  0.0979  -0.0120  -0.0263
 0.0979  0.0255  -0.0022  -0.0092
-0.0120 -0.0022   0.0025  -0.0030
-0.0263 -0.0092  -0.0030   0.2984

-------------------------------------------
-------------------------------------------
---------- Correlation Matrix -------------
-------------------------------------------
 1.0000  0.9905 -0.3915 -0.0777
 0.9905  1.0000 -0.2722 -0.1049
-0.3915 -0.2722  1.0000 -0.1106
-0.0777 -0.1049 -0.1106  1.0000

-------------------------------------------
```

7.5 Quality of fit and discrimination among rival models

Another important topic when experimental data must be interpreted through a mathematical model is related to the quality of fit or, in other words, how accurately the model describes experimental data. There is the possibility to calculate a quantitative representation of the goodness of experimental data fitting. As a first approach, we can adopt the well-known parameter R^2 (correlation coefficient or determination coefficient) that is a measure of data description quality and is defined as follows:

$$R^2 = 1 - \frac{\sum_{i=1}^{n} (y_i - \hat{y}_i)^2}{\sum_{i=1}^{n} (y_i - \bar{y})^2} \tag{7.20}$$

In this expression, $\bar{y}$ represents the numerical average of the experimental data. A value of R^2 close to 1 indicates generally a good quality of fit. Anyway, the only use of R^2 could be not enough to judge if a model is adequate for the description of experimental data as this parameter does not contain information about the number of adjustable parameters that are included in the model. Another situation in which the use of correlation coefficient is not recommended, is the issue related to the selection of a model between different candidates.

More suitable, at these purposes, is the "adjusted" R^2 (R^2_{adj}) defined as follows:

$$R^2_{adj} = 1 - \frac{(1 - R^2)(n - 1)}{(n - p - 1)} \tag{7.21}$$

This coefficient, in this form, is sensitive to the number of adjustable parameter (p) and is always lower than R^2. With adjusted R^2_{adj} it is possible to compare, on statistical basis, different models with different numbers of parameters.

With the purpose of model comparison, selection or discrimination, many other statistical tests have been proposed in the specialized books and literature and, even if a deep discussion on these tests is outside of the scope of this book, some practical formulas are reported as follows:

First, we need to define three standard quantities that are functions of experimental and calculated y data:

$$\text{RSS} = \sum_{i=1}^{n} (y_i - \hat{y}_i)^2 \tag{7.22}$$

$$\text{TSS} = \sum_{i=1}^{n} (y_i - \bar{y})^2 \tag{7.23}$$

$$\text{ESS} = \sum_{i=1}^{n} (\hat{y}_i - \bar{y})^2 \tag{7.24}$$

The most used expressions for statistical tests are the following:

$$F = \frac{(R^2/p)}{(1 - R^2)/(n - p - 1)} \quad \text{Fisher's test} \tag{7.25}$$

$$F_E = \frac{(\text{TSS} - \text{RSS})/p}{(\text{RSS})/(n-p)} \quad \text{Fisher's E- test} \tag{7.26}$$

$$F_m = \frac{\text{ESS}/(p-1)}{(\text{RSS})/(n-p)} \quad \text{Modified Fisher's test} \tag{7.27}$$

In all the cases, the values of the statistical tests from expressions (7.25) to (7.27) should be compared with tabulated reference values for assessing the statistical significance of a single model but, when used for comparing different models, the highest calculated value of the test should correspond to the best model. Obviously, this is only a statistical inference and the choice among rival models must be made, also, by taking into account for parameter physical meaning.

7.6 Sensitivity analysis

The sensitivity analysis aims to evaluate how the parameters influence the model response or, in other words, how the parameter uncertainty affects the model prediction. As a first approach, the parametric sensitivity can be expressed by jacobian matrix defined as

$$J = \begin{vmatrix} \partial\hat{y}_1/\partial\beta_1 & \partial\hat{y}_1/\partial\beta_2 & \dots & \partial\hat{y}_1/\partial\beta_p \\ \partial\hat{y}_2/\partial\beta_1 & \partial\hat{y}_2/\partial\beta_2 & \dots & \partial\hat{y}_2/\partial\beta_p \\ \dots & \dots & & \dots \\ \partial\hat{y}_{Ns}/\partial\beta_1 & \partial\hat{y}_{Ns}/\partial\beta_2 & \dots & \partial\hat{y}_{Ns}/\partial\beta_p \end{vmatrix} \tag{7.28}$$

Or in a more compact form:

$$J_{i,j} = \partial\hat{y}_i/\partial\beta_j \quad i = 1 \dots Ns \quad j = 1 \dots p \tag{7.29}$$

This matrix represents the dependency of the model from each parameter, in correspondence of each experimental point. In this form, the evaluation of parametric sensitivity is not very useful. However, the presence of high values for some elements of the jacobian matrix reveals high sensitivity of the model with respect to that parameter.

An alternative approach for a sensitivity study consists of plotting the objective function versus each parameter by fixing all other parameters at their optimum value. This approach will be better explained in a solved example.

7.7 Residual analysis

When a mathematical model is applied to the statistical description of experimental data, this description is not perfect and a deviation (residual) is expected between the experimental measurements and model prediction. This deviation can de expressed as follows:

$$\varepsilon_i = y_i - \hat{y}_i \tag{7.30}$$

The analysis of these residuals represents a further useful tool to assess the statistical validity of the model. In fact, if the model has a good statistical behavior, the residuals must be randomly distributed around zero without any recognizable patterns (e.g., a group of negative residuals followed by a group of positive residuals). This type of analysis can be performed, for example, by plotting the residuals as a function of the independent variable X; such a plot would be of the type reported in Figure 7.3.

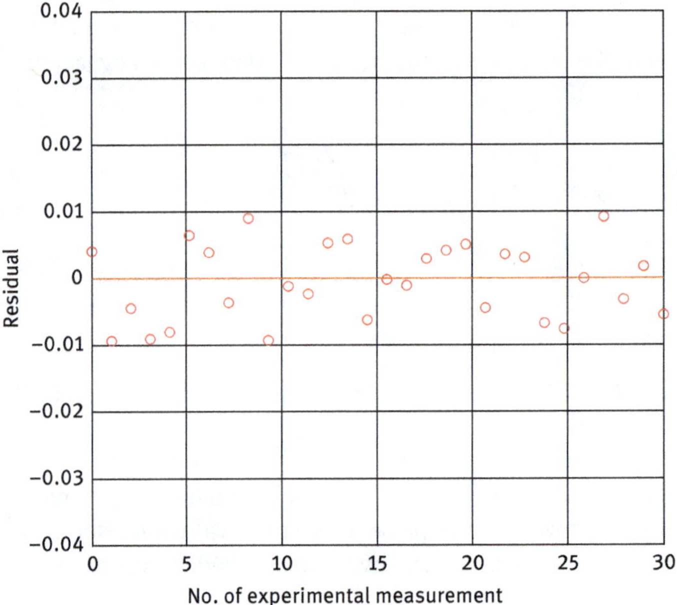

Figure 7.3: Example of residual plot.

In this way, the evaluation of the random distribution of the residuals can be judged only graphically and not from an objective point of view. The alternative consists of building a residual normal probability plot such that reported in Figure 7.4.

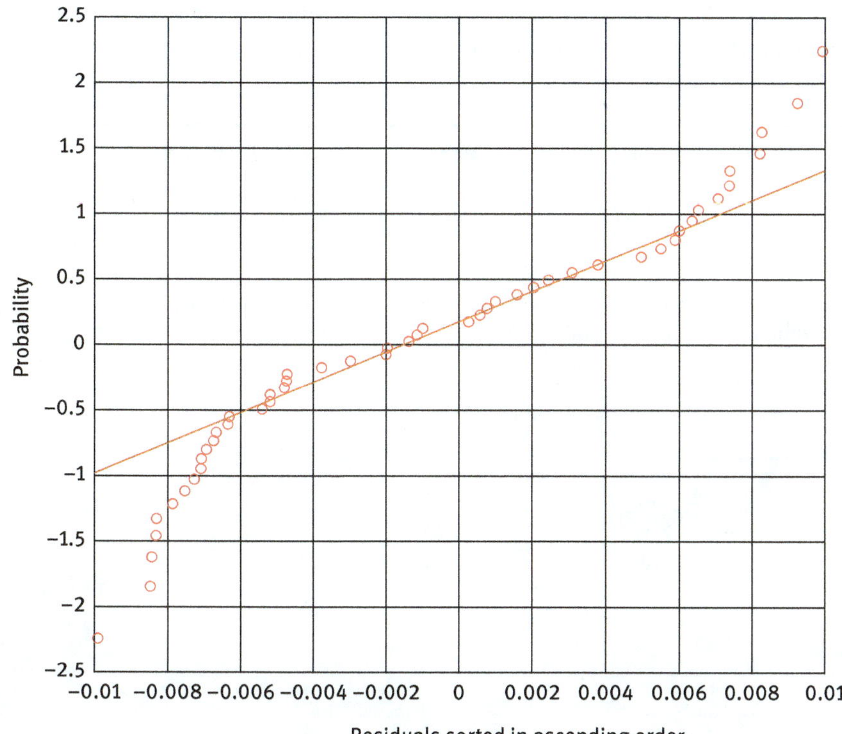

Figure 7.4: Example of residual normal probability plot.

The construction of the normal probability plot proceeds through few simple steps as briefly summarized here:

- Sort the residuals in ascending order
- Calculate the following quantity for each residual (i is the position of the residual in the ordered list):

$$f_i = \frac{i - 0.375}{n + 0.25} \tag{7.31}$$

  Where n is the total number of residuals
- Calculate the score of each residual according to the standardized normal distribution:

$$\Phi(z) = \frac{1}{\sqrt{2\pi}} e^{-\frac{1}{2}z^2} \tag{7.32}$$

  The score of the residual i is then calculated as follows:

$$s_i = \Phi^{-1}(\mathrm{f_i}) \tag{7.33}$$

The normal probability plot is finally obtained by plotting s_i as a function of the sorted residuals. If the data in this plot are aligned along a straight line, the distribution of the residuals is normal.

Example 7.3 Equilibrium reaction, liquid-phase isothermal batch reactor, parameter fitting, residuals analysis

In a batch reactor, a single equilibrium reaction occurs as in the following scheme:

$$A + B \longleftrightarrow C + D$$

This reaction was studied by collecting percentage conversion of A as a function of time. The experimental observations are reported in Table 7.4.

Table 7.4: Experimental data conversion versus time, Example 7.3.

Time (s)	X_A (–)
0	0
10	7
20	12
30	19
45	25
60	30
75	38
90	41
120	48
150	52
180	57
210	60
240	64
270	66
300	68
400	70
500	70
600	70

The expression for the reaction rate is defined as follows:

$$r = k_C\left(C_A C_B - \frac{1}{K_E} C_C C_D\right)$$

The initial concentrations in the reactor were $C_A = 1{,}000$ mol/m^3, $C_B = 5{,}000$ mol/m^3, $C_C = 0$ mol/m^3 and $C_D = 0$ mol/m^3. The material balance can be written for each component as follows:

$$\frac{dC_i}{dt} = v_i r \quad i = A, B, C, D \quad v_i = [-1, -1, +1, +1]$$

The purpose of the present solved example is to evaluate by fitting the two adjustable parameters (kinetic and equilibrium constants) and the related confidence intervals, covariance and correlation matrix. Perform, at last, the residual analysis by building the residual plot and the normal probability plot for the residuals.

Matlab code for the solution of this example and results are reported as follows Figure 7.5:

```
% example 7.3
clear, clc
%% Reaction scheme
% A + B <-> C + D
%% Experimental data
TimeEXP=[0; 10; 20; 30; 45; 60; 75; 90; 120; 150; 180; 210; 240; 270; 300; 400; 500; 600];
XAEXP= [0; 7; 12; 19; 25; 30; 38; 41; 48; 52; 57; 60; 64; 66; 68; 70; 70; 70];
%% Test conditions
cA0=1000;    % mol/m3
cB0=5000;    % mol/m3
cC0=0;       % mol/m3
cD0=0;       % mol/m3
c0=[cA0,cB0,cC0,cD0];

%% Parameter estimation
kc=1e-6;    % m3/mol/s
KE=0.5;     % -
k0=[kc KE];
lb=[0 0] ;
ub=[1e-5 1] ;
options=optimoptions(@lsqnonlin,'Display','Iter');
f1 = @(k)obj_ex8_03(k,c0,TimeEXP,XAEXP);
[k,resnorm,residual,exitflag,output,lambda,jacobian]=lsqnonlin(f1,k0,lb,ub,
options);
ci = nlparci(k,residual,'jacobian',jacobian);
CIkc=abs(ci(1,1)-ci(1,2))/2;
CIKE=abs(ci(2,1)-ci(2,2))/2;
Jacobian=full(jacobian);
s2=resnorm/length(TimeEXP);
```

```
covarM=s2*inv(Jacobian'*Jacobian) ;
corrM=corrcov(covarM) ;
stdev=sqrt(diag(covarM));

%% Simulation
tspan=linspace(0,TimeEXP(end),1e3);
f2=@(k,tspan)model_ex8_03(k,tspan,c0);
[XApred,delta]=nlpredci(f2,tspan,k,residual,'Jacobian',jacobian);
f3=@(T,c)ode_ex8_03(T,c,k);
[T, c]=ode23s(f3,tspan,c0);
cA=c(:,1);
XA=100*(1-cA./c0(1));

%% Output
subplot(1,3,1)
plot(TimeEXP,XAEXP,'sb')
hold on
plot(tspan,XA,'-b',tspan,XApred+delta,'-b',tspan,XApred-delta,'-b')
hold off
xlabel('\it t \rm [s]'), ylabel('\it X_{A} \rm [%]')
grid on

subplot(1,3,2)
plot(TimeEXP,residual,'ro')
grid
xlabel('\it t \rm [s]'), ylabel('\it residual \rm [%]')

nres=length(residual);
sortres=sort(residual);
ires=linspace(1,nres,nres);
fres=(ires-0.375)/(nres+0.25);
prob=-sqrt(2)*erfcinv(2*fres);
subplot(1,3,3)
plot(sortres,prob,'ro')
grid
xlabel('\it residual \rm [%]'), ylabel('\it probability \rm [-]')

disp('———————————————————————————————————————-')
disp('———————- Fitted Parameters ————————————-')
disp('———————————————————————————————————————-')
disp(['kc =', num2str(k(1)), ' +/-', num2str(CIkc), ' m3/mol/s'])
disp(['KE =', num2str(k(2)), ' +/-', num2str(CIKE), '-'])
disp('———————————————————————————————————————-')

%% Output statistics
str = strvcat('kc','KE') ;
disp('———————————————————————————————————————-')
```

```
disp('------------ Covariance Matrix ----------------------')
disp('-----------------------------------------------------')
disp(' kc KE')
disp([str, num2str(covarM)]) ;
disp('-----------------------------------------------------')

disp('-----------------------------------------------------')
disp('------------ Correlation Matrix ----------------------')
disp('-----------------------------------------------------')
disp(' kc KE')
disp([str, num2str(corrM)]) ;
disp('-----------------------------------------------------')

function [ df ] = obj_ex8_03(k,c0,TimeEXP,XAEXP)
%% ODE function
   f=@(T,c)ode_ex8_03(T,c,k);
   options=odeset('RelTol',1e-8);
   [T, c]=ode23s(f,TimeEXP,c0,options);
   cA=c(:,1);
   XA=100*(1-cA./c0(1));
df=XA-XAEXP;
end

function [ XA ] = model_ex8_03(k,tspan,c0)
%% ODE function
f=@(T,c)ode_ex8_03(T,c,k);
options=odeset('RelTol',1e-8);
[T, c]=ode23s(f,tspan,c0,options);
cA=c(:,1);
XA=100*(1-cA./c0(1));
end

function dcdt = ode_ex8_03(T,c,k)
% Kinetic parameters
kc = k(1) ;
KE = k(2) ;
% Concentrations
cA=c(1); cB=c(2); cC=c(3); cD=c(4);
% Reaction rate
r=kc*(cA*cB-(1/KE)*cC*cD); % mol/m3/s
% ODE system
dcdt(1)=-r;
dcdt(2)=-r;
dcdt(3)=+r;
dcdt(4)=+r;
dcdt=dcdt(:);
end
```

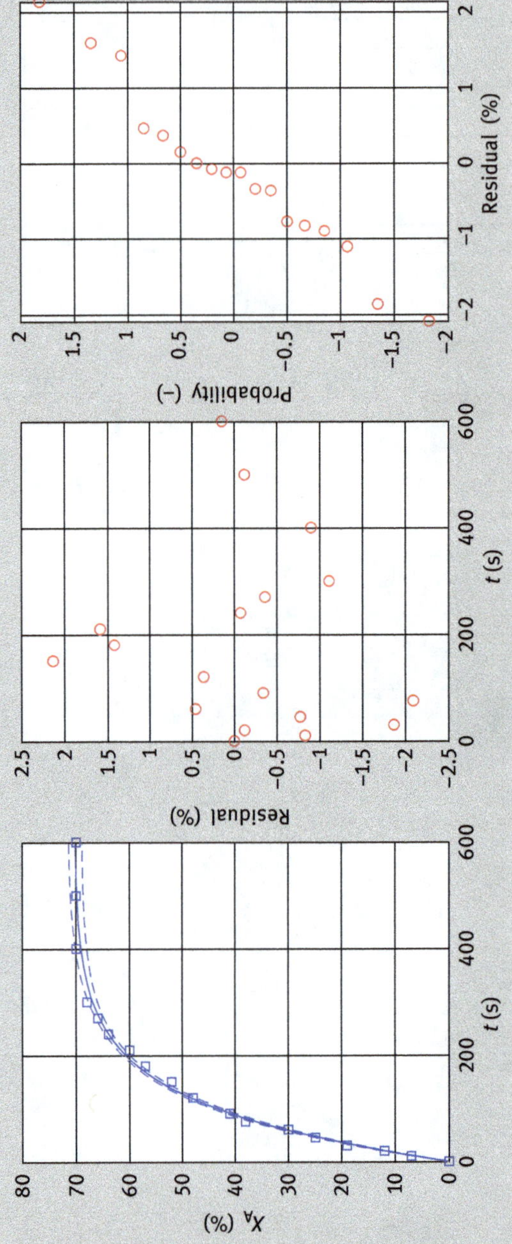

Figure 7.5: (Left) Experimental and calculated conversion of A versus time; (center) residuals versus time; (right) normal probability plot of the residuals.

```
                         Norm of    First-order
Iteration Func-count  f(x)    step      optimality
     0       3     307.373              1.1e+04
     1       6     59.9681    0.223637    2.57e+03
     2       9     21.2115    0.04716      80.9
     3      12     20.9102   0.00839471      0.939
     4      15     20.9095  0.000355947      0.0225
     5      18     20.9095  9.34143e-06    0.000782
------------------------------------------
------------ Fitted Parameters -------------
------------------------------------------
kc =1.2837e-06 +/-4.7445e-08 m3/mol/s
KE =0.38702 +/-0.032044-
------------------------------------------
------------ Covariance Matrix -------------
------------------------------------------
kc       KE
kc 4.4524e-16 -1.3286e-10
KE-1.3286e-10  0.0002031
------------------------------------------
------------ Correlation Matrix ------------
------------------------------------------
   kc        KE
kc    1    -0.44182
KE-0.44182       1
------------------------------------------
```

Example 7.4 Single reaction, liquid-phase isothermal batch reactor, multiple kinetic runs, parameter fitting, residuals analysis, surface of objective function

In a batch reactor, a single reaction occurs as in the following scheme:

$$A + B \rightarrow C$$

This reaction was studied by collecting fractional conversion of A as a function of time at different temperatures. The experimental observations are reported in Table 7.5.

Table 7.5: Experimental data conversion versus time at different temperatures, Example 7.4.

$T = 100$ °C		$T = 120$ °C		$T = 160$ °C	
t (s)	X_A (–)	t (s)	X_A (–)	t (s)	X_A (–)
0	0.000	0	0.000	0	0.000
11	0.090	8	0.100	4	0.178
45	0.298	35	0.431	7	0.317

Table 7.5 (continued)

$T = 100$ °C		$T = 120$ °C		$T = 160$ °C	
t (s)	X_A (–)	t (s)	X_A (–)	t (s)	X_A (–)
100	0.526	82	0.648	13	0.431
145	0.626	220	0.855	21	0.530
170	0.672	289	0.907	28	0.641
219	0.730			42	0.737
268	0.779			65	0.845
				136	0.944
				232	0.980
				288	0.991

The reaction rate is defined as in the following expression:

$$r = k_C C_A C_B$$

The initial concentrations in the reactor were $C_A = 0.20$ mol/m^3, $C_B = 0.25$ mol/m^{3}and $C_C = 0$ mol/m^3. The material balance can be written for each component as follows:

$$\frac{dC_i}{dt} = v_i r \; i = \text{A, B, C} \; v_i = [-1, -1, +1]$$

The runs reported in the previous table are performed at different temperature and therefore the Arrhenius equation must be used for taking into account the dependence of kinetic constant from temperature as follows:

$$k = k_{\text{ref}} e^{\left[\frac{Ea}{R}\left(\frac{1}{T_{\text{ref}}} - \frac{1}{T}\right)\right]}$$

The reference temperature is fixed at $T = 110$ °C.

The purpose of the present solved example is to evaluate, by simultaneous fitting on the three runs, the two adjustable parameters (reference kinetic constant and activation energy) and the related confidence intervals, covariance and correlation matrix. Perform, at last, the residual analysis by build the residual plot and parity plot.

In this particular example, as the problem involves the determination of two adjustable parameters, a further possible plot can be drawn that represents a contour plot of the objective function as a function of the two parameters.

Matlab code for the solution of this example and results are reported as follows Figure 7.6:

```
% example 7.4
clc,clear
%% experimental data
t1=[0 11 45 100 145 170 219 268];
x1=[0 0.09 0.298 0.526 0.626 0.672 0.730 0.779];
t2=[0 8 35 82 220 289];
x2=[0 0.100 0.431 0.648 0.855 0.907];
t3=[0 4 7 13 21 28 42 65 136 232 288];
x3=[0 0.178 0.317 0.431 0.530 0.641 0.737 0.845 0.944 0.98 0.991];
data={t1' x1' t2' x2' t3' x3'};
Nrun = 3;
Trun = [100 120 160];
for j=1:Nrun
 Ndata(j)=length(data{j*2});
end

%% parameter estimation
C0 = [0.20 0.25 0.00];
Kref0= 0.05;
Tref = 110+273.15;
Ea0 = 39000;
k0 = [Kref0 Ea0]; % initial kinetic parameters
lb = [ 0 0];          % lower bound
ub = [10 100000]; % upper bound
options=optimoptions(@lsqnonlin,'Display','Iter','MaxFunctionEvaluations',1000);
objf=@(kp)obj_ex8_04(kp,data,C0,Tref,Nrun,Trun);
[k,resnorm,residual,exitflag,output,lambda,jacobian]=lsqnonlin(objf,k0,lb,ub,
options);
ci = nlparci(k,residual,'jacobian',jacobian);
CIkc=abs(ci(1,1)-ci(1,2))/2;
CIEa=abs(ci(2,1)-ci(2,2))/2;
Jacobian=full(jacobian);
s2=resnorm/length(residual);
covarM=s2*inv(Jacobian'*Jacobian) ;
corrM=corrcov(covarM) ;
stdev=sqrt(diag(covarM));

%% output parameters and statistics
disp('------------------------------')
disp('--------- Fitted Parameters ---------')
disp('------------------------------')
fprintf(' Parameter 1 - Kref : %10.4f C.I.: +/- %10.4f \n',k(1),CIkc)
fprintf(' Parameter 2 - Ea : %10.1f C.I.: +/- %10.1f \n',k(2),CIEa)
disp('------------------------------')
disp('--------------------------')
disp('------ Covariance Matrix -------')
```

```
disp('-----------------------')
disp(covarM);
disp('-----------------------')
disp('------ Correlation Matrix -------')
disp('-----------------------')
disp(corrM);
disp('-----------------------')

%% simulation
tspan = 0:0.1:300;
Kref = k(1);
Ea = k(2);
Rgas = 8.314;
for j=1:Nrun
  Tk=Trun(j)+273.15;
  kc = Kref*exp(Ea/Rgas*(1/Tref-1/Tk));
  ode=@(t,C)ode_ex8_04(t,C,kc);
  [tc,Cc]=ode15s(ode,tspan,C0);
  x=(C0(1)-Cc(:,1))./C0(1);
  Xa(:,j)=x;
  jrun=j;
  modelf=@(k,tspan)model_ex8_04(k,tspan,data,C0,Tref,Nrun,Trun,jrun);
  [XApred,delta] = nlpredci(modelf,tspan,k,residual,'Jacobian',
  jacobian);
  XXp(:,j)=XApred+delta;
  XXm(:,j)=XApred-delta;
end

%% plots
subplot(2,2,1)
plot(t1,x1,'ro',t2,x2,'bo',t3,x3,'ko', . . .
  tc,Xa(:,1),'-r',tc,Xa(:,2),'-b',tc,Xa(:,3),'-k', . . .
  tc,XXp(:,1),'r-',tc,XXm(:,1),'r-', . . .
  tc,XXp(:,2),'b-',tc,XXm(:,2),'b-', . . .
  tc,XXp(:,3),'k-',tc,XXm(:,3),'k-')
grid
xlabel('Time (s)')
ylabel('Xa (-)')
legend('T=100°C','T=120°C','T=160°C')
subplot(2,2,2)
plot(residual,'-ro')
grid
xlabel('Experiment number')
ylabel('Residual (-)')
subplot(2,2,3)
xS=[];
for j=1:Nrun
  xS=[xS;data{2*j}];
end
```

```
xC=xS-residual;
plot(xC,xS,'ro',[0 1],[0 1],'-b')
grid
xlabel('Calculated conversion of A (-)')
ylabel('Experimental conversion of A (-)')
subplot(2,2,4)
Nval=200;
Krx=linspace(0.05,0.06,Nval);
Eay=linspace(37500,38500,Nval);
[X,Y] = meshgrid(Krx,Eay);
Lx=length(Krx);
Ly=length(Eay);
for jx=1:Lx
  for jy=1:Ly
    kkk(1)=Krx(jx);
    kkk(2)=Eay(jy);
    Z(jx,jy)= error_ex8_04(kkk,data,C0,Tref,Nrun,Trun);
  end
end
contourf(X,Y,Z,Nval,'ShowText','off')
c = colorbar;
c.Label.String = 'Objective function (-)';
xlabel('Kref (m3/(s mol))')
ylabel('Ea (J/mol)')
hold on
plot(k(1),k(2),'yo')

function [res] = obj_ex8_04(kp,data,C0,Tref,Nrun,Trun)
Kref = kp(1);
Ea = kp(2);
R = 8.314;
res=[];
for j=1:Nrun
  Tk=Trun(j)+273.15;
  kc = Kref*exp(Ea/R*(1/Tref-1/Tk));
  ts = data{j*2-1};
  xs = data{j*2};
  tspan = ts; % time range for integration (s)
  ode=@(t,C)ode_ex8_04(t,C,kc);
  [tc,Cc]=ode15s(ode,tspan,C0);
  xa=(C0(1)-Cc(:,1))./C0(1);
  resx = xs-xa;
  res=[res;resx];
end
end

function [Xc] = model_ex8_04(kp,Xs,data,C0,Tref,Nrun,Trun,jrun)
Kref = kp(1);
Ea   = kp(2);
```

```
R    = 8.314;
Tk   = Trun(jrun)+273.15;
kc   = Kref*exp(Ea/R*(1/Tref-1/Tk));
tspan = Xs;         % time range for integration (s)
ode=@(t,C)ode_ex8_04(t,C,kc);
[tc,Cc]=ode15s(ode,tspan,C0);
xa=(C0(1)-Cc(:,1))./C0(1);
Xc=xa;
end

function [dC] = ode_ex8_04(t,C,k)
r=k*C(1)*C(2);
dC(1)=-r;
dC(2)=-r;
dC(3)=+r;
dC=dC';
end

function [er] = error_ex8_04(kp,data,C0,Tref,Nrun,Trun)
[res] = obj_ex8_04(kp,data,C0,Tref,Nrun,Trun);
Nres=length(res);
er=sum(res.^2);
end

                          Norm of    First-order
Iteration  Func-count   f(x)       step       optimality
    0        3     0.0103522                  14.9
    1        6    0.00437611     3.63931       0.66
    2        9    0.00436276    0.0106602    0.000558
    3       12    0.00436276    0.00525262    1.85e-06
------------------------------------------------
---------------- Fitted Parameters -----------------
------------------------------------------------
Parameter 1 - Kref :   0.0546  C.I.: +/-   0.0019
Parameter 2 - Ea  :  38099.8  C.I.: +/-   1439.7
------------------------------------------------
---------------- Covariance Matrix ----------------
------------------------------------------------
1.0e+05 *
0.0000  -0.0000
-0.0000  4.4561
------------------------------------------------
-------------- Correlation Matrix -----------------
------------------------------------------------
1.0000  -0.5386
-0.5386  1.0000
------------------------------------------------
```

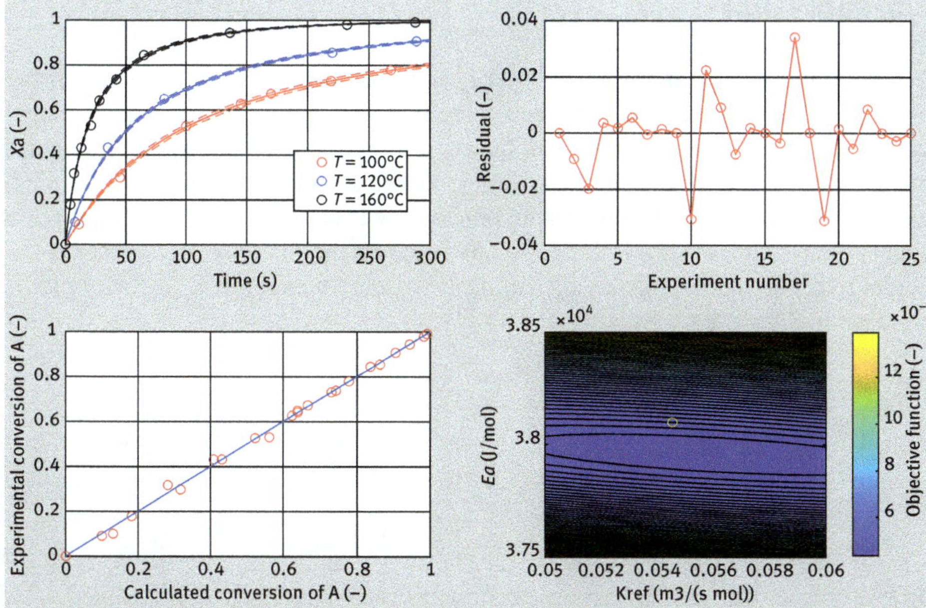

Figure 7.6: Experimental and calculated conversion of A at different temperature (up-left); (up-right) residuals plot; (down-left) parity plot of the fit; (down-right) contour map of the objective function. *Ea* as a function of k_{ref}. Color map represents the value of the objective function.

7.8 Parity plot

A very useful way to represent the performance of a mathematical model in the description of experimental data, is represented by the parity plot, used in particular when many experimental runs must be descripted together with the same model. This plot is constructed in a very simple way by reporting experimental values as a function of the calculated ones or the inverse. The obtained plot has a square shape and the more the points are near to the diagonal, the closer is the model to the experiments and the fitting is better. An example of a parity plot is reported in Figure 7.7.

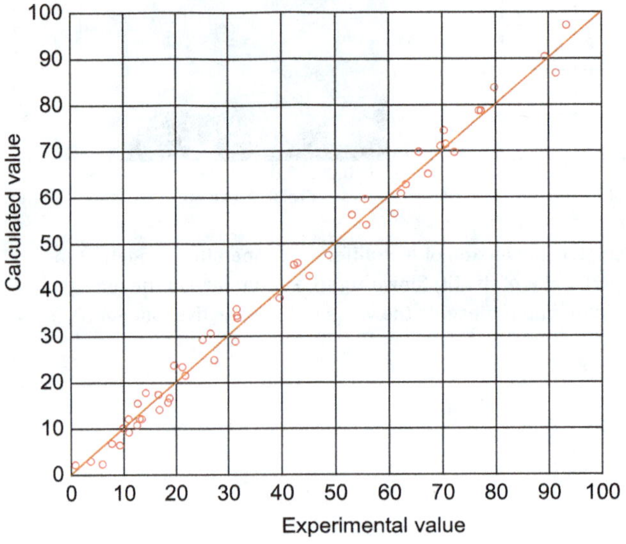

Figure 7.7: Example of parity plot.

Example 7.5 Single reaction, isothermal plug flow tubular reactor, runs at different temperatures parameter fitting, parity plot

In tubular plug-flow continuous reactor, a single reaction occurs as in the following scheme:

$$A + B \rightarrow C$$

This reaction was studied by collecting fractional conversion of A in stationary conditions at different temperature. For each run only a value of conversion is available. The experimental observations are reported in Table 7.6.

Table 7.6: Experimental data conversion versus temperature, Example 7.5.

T (°C)	X_A (–)
160	0.002
180	0.005
200	0.02
220	0.03
240	0.12
260	0.32
280	0.49
300	0.85
320	0.94
360	0.99
380	0.99

The expression for the reaction rate is defined as in the following expression:

$$r = k_C C_A C_B$$

The feed molar flowrates to the reactor were $F_A = 20$ mol/s, $F_B = 25$ mol/s and $F_C = 0$ mol/s. The reactor was charged with 50 kg of catalyst and fed with a total volumetric flowrate of 0.25 m^3/s. The material balance can be written for each component as follows:

$$\frac{dF_i}{dw} = v_i r \quad i = \text{A, B, C} \quad v_i = [-1, -1, +1]$$

The runs reported in the previous table are performed at different temperature and therefore the Arrhenius equation must be used for taking into account the dependence of kinetic constant from temperature as follows:

$$k = k_{ref} e^{\left[\frac{Ea}{R}\left(\frac{1}{T_{ref}} - \frac{1}{T}\right)\right]}$$

The reference temperature is fixed at $T = 250$ °C.

The purpose of the present solved example is to evaluate, by simultaneous fitting on the runs at different temperatures, the two adjustable parameters (reference kinetic constant and activation energy) and the related confidence intervals, covariance and correlation matrix. Perform, at last, the residual analysis by building a parity plot.

Matlab code for the solution of this example and results are reported as follows Figure 7.8:

```
% example 7.5
clc,clear
%% experimental data
Ts = [160 180 200 220 240 260 280 300 320 360 380];
xAs = [0.002 0.005 0.02 0.03 0.12 0.32 0.49 0.85 0.94 0.99 0.99];
data = [Ts' xAs'];
F0=[20 25 0];
Wcat = 50;
Nrun=length(Ts);
Qtot = 0.25;

%% parameter estimation
Kref0= 0.0005;
Tref = 250+273.15;
Ea0 = 39000;
k0 = [Kref0 Ea0]; % initial kinetic parameters
lb = [ 0 0];        % lower bound
ub = [10 1e7];      % upper bound
options=optimoptions(@lsqnonlin,'Display','Iter','MaxFunctionEvaluations',1000);
objf=@(kp)obj_ex8_05(kp,data,F0,Tref,Nrun,Wcat,Qtot);
[k,resnorm,residual,exitflag,output,lambda,jacobian]=lsqnonlin(objf,k0,lb,ub,
options);
ci = nlparci(k,residual,'jacobian',jacobian);
CIkc=abs(ci(1,1)-ci(1,2))/2;
CIEa=abs(ci(2,1)-ci(2,2))/2;
Jacobian=full(jacobian);
s2=resnorm/length(residual);
covarM=s2*inv(Jacobian'*Jacobian) ;
corrM=corrcov(covarM) ;
stdev=sqrt(diag(covarM));

%% output parameters and statistics
disp('--------------------------------')
disp('---------- Fitted Parameters ----------')
disp('--------------------------------')
fprintf(' Parameter 1 - Kref : %10.4e C.I.: +/- %10.4f \n',k(1),CIkc)
fprintf(' Parameter 2 - Ea : %10.1f C.I.: +/- %10.1f \n',k(2),CIEa)
disp('--------------------------------')
disp('--------------------------')
disp('------- Covariance Matrix --------')
disp('--------------------------')
disp(covarM) ;
disp('--------------------------')
disp('------- Correlation Matrix --------')
disp('--------------------------')
disp(corrM) ;
disp('--------------------------')
```

```
%% simulation
Tx=180:380;
Kref = k(1);
Ea = k(2);
Rgas = 8.314;
wspan=0:Wcat/1000:Wcat;
for j=1:length(Tx)
  Txk=Tx(j)+273.15;
  kc = Kref*exp(Ea/Rgas*(1/Tref-1/Txk));
  ode=@(w,F)ode_ex8_05(w,F,kc,Qtot);
  [wc,Fc]=ode15s(ode,wspan,F0);
  x=(F0(1)-Fc(end,1))./F0(1);
  Xa(j)=x;
end

%% plots
subplot(1,2,1)
plot(Ts,xAs,'ro',Tx,Xa)
grid
xlabel('Temperature (°C)')
ylabel('Xa (-)')

subplot(1,2,2)
res = obj_ex8_05(k,data,F0,Tref,Nrun,Wcat,Qtot);
xAc=res+xAs';
plot(xAc,xAs,'ro',[0 1],[0 1])
grid
xlabel('Experimental conversion (-)')
ylabel('Calculated conversion (-)')

function [res] = obj_ex8_05(kp,data,F0,Tref,Nrun,Wcat,Qtot)
Kref = kp(1);
Ea = kp(2);
R = 8.314;
Ts = data(:,1);
xAs = data(:,2);
for j=1:Nrun
  Tk=Ts(j)+273.15;
  kc = Kref*exp(Ea/R*(1/Tref-1/Tk));
  wspan=0:Wcat/1000:Wcat;
  ode=@(w,F)ode_ex8_05(w,F,kc,Qtot);
  [wc,Fc]=ode15s(ode,wspan,F0);
  xa(j)=(F0(1)-Fc(end,1))./F0(1);
end
res=xa'-xAs;
end
```

```
function [dF] = ode_ex8_05(w,F,k,Qtot)
%% ——————————-
% example 7.5 - differential equations
%———————————-
C=F/Qtot;
r=k*C(1)*C(2);
dF(1)=-r;
dF(2)=-r;
dF(3)=+r;
dF=dF';
end
```

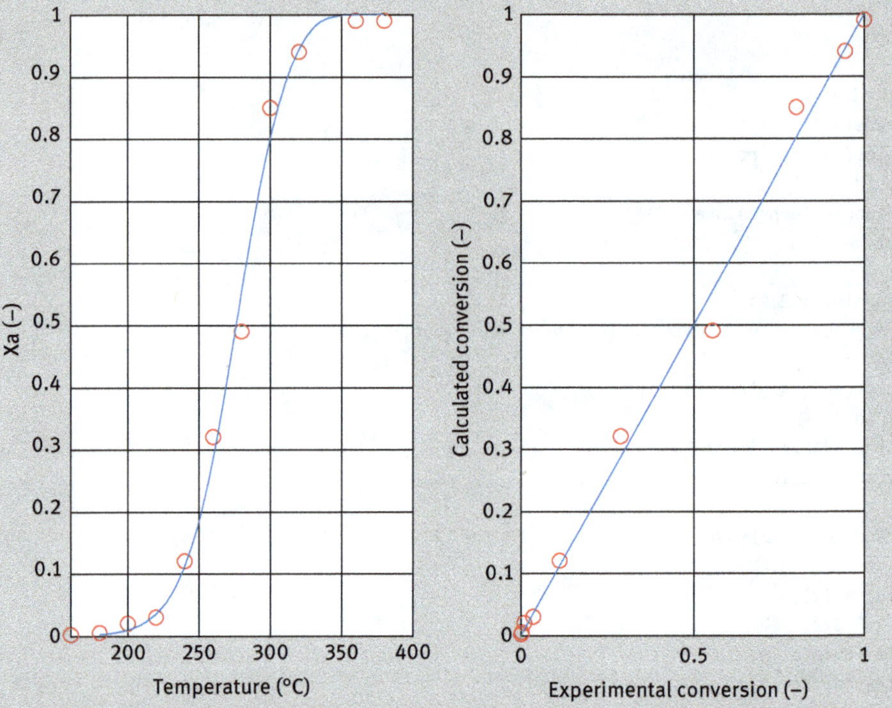

Figure 7.8: Experimental and calculated conversion of A versus temperature, dots are experimental values, continuous curves represent model (left); (right) parity plot of the fit.

```
                              Norm of     First-order
Iteration Func-count  f(x)       step        optimality
    0        3       3.8837                 195
    1        6       1.72044       10         73
    2        9      0.826921     11.516        9.36
    3       12      0.238894     2.00929       0.21
    4       15      0.0304889       20       0.0597
    5       18      0.0111613    23.9151       1.47
```

```
   6     21   0.00808385     4.01952    1.11e+03
   7     24   0.0080088    0.740714    0.000128
   8     27   0.00800866    0.145107     0.00184

------------------------------------------------
---------------- Fitted Parameters ----------------
------------------------------------------------
Parameter 1 - Kref : 1.1073e-05  C.I.: +/-   0.0000
Parameter 2 - Ea   :  129260.7  C.I.: +/-  19841.3
------------------------------------------------
------------------------------------------------
---------------- Covariance Matrix ---------------
------------------------------------------------
1.0e+07 *
0.0000  -0.0000
-0.0000  6.2942
------------------------------------------------
-------------- Correlation Matrix ----------------
------------------------------------------------
1.0000  -0.8365
-0.8365  1.0000
------------------------------------------------
```

Example 7.6 Single reaction, liquid-phase isothermal batch reactor, Langmuir–Hinshelwood model, particle swarm optimization

In a batch reactor a single reaction occurs as in the following scheme:

$$A + B \rightarrow C$$

The reaction is catalyzed by a solid and was studied by collecting concentration data of A, B and C as a function of time. The experimental observations are reported in Table 7.7.

Table 7.7: Experimental data concentration versus time, Example 7.6.

Time (s)	C_A (mol/m^3)	C_B (mol/m^3)	C_C (mol/m^3)
0	75	50	0
10	65	39	10
20	55	31	18
30	47	22	27
40	41	16	33
60	33	8	42
80	28	6	47
120	26	2	48
160	24	1	48
180	25	0.5	50

The expression for the reaction rate is defined according to the well-known Langmuir–Hinshelwood reaction–adsorption model:

$$r = \frac{kK_A C_A K_B C_B}{(1 + K_A C_A + K_B C_B + K_C C_C)^2}$$

The initial concentrations in the reactor were, respectively, for A, B and C: $C_A = 75$ mol/m^3, $C_B = 50$ mol/m^3 and $C_C = 0$ mol/m^3. The material balance can be written for each component as follows:

$$\frac{dC_i}{dt} = v_i r \quad i = \mathrm{A, B, C} \quad v_i = [-1, -1, +1]$$

The purpose of the present example is to evaluate, by fitting, the four adjustable parameters (one kinetic constant and three adsorption equilibrium constants). For illustrative purpose, the fitting is performed with an evolutionary algorithm like PSO.

Matlab code for the solution of this example and results are reported as follows Figure 7.9:

```
% example 7.6
clc,clear
data=[ 0 75 50 0
    10 65 39 10
    20 55 31 18
    30 47 22 27
    40 41 16 33
    60 33 8 42
    80 28 6 47
    120 26 2 48
    160 24 1 48
    180 25 0.5 50 ];

ts = data(:,1);
cas = data(:,2);
cbs = data(:,3);
ccs = data(:,4);

%% parameter estimation
c0=[75 50 0];
objf=@(kp)obj_ex8_06(kp,data,c0);
k0 = [7.5 0.022 0.035 0.007]; % initial kinetic constants
lb = [ 0   0   0   0 ];
ub = [10   1   1   1 ];
nvars=4;
options=optimoptions(@particleswarm,'Display','Iter','SwarmSize',30);
k=particleswarm(objf,nvars,lb,ub,options);
disp('--------------------------')
disp('------ Fitted Parameters -------')
disp('--------------------------')
```

```
for j=1:nvars
  fprintf(' Parameter k %1.0f = %10.5e \n',j,k(j))
end

%% simulation
tspan=0:0.1:200; % time range for integration (s)
ode=@(t,C)ode_ex8_06(t,C,k);
[tx,cx]=ode15s(ode,tspan,c0);

%% plot
plot(tx,cx(:,1),'r-',tx,cx(:,2),'k-',tx,cx(:,3),'b-', . . .
    ts,cas,'ro',ts,cbs,'ko',ts,ccs,'bo')
grid
xlabel('Time (s)')
ylabel('Concentration (mol/m3)')
legend('A','B','C')

function [fobj] = obj_ex7_06(kp,data,C0)
ts = data(:,1);
cas = data(:,2);
cbs = data(:,3);
ccs = data(:,4);
ndat= length(ts);
tspan = ts;          % time range for integration (s)
ode=@(t,C)ode_ex8_06(t,C,kp);
[tc,cc]=ode15s(ode,tspan,C0);
fobj=0;
for j=1:ndat
   fobj = fobj + (cas(j)-cc(j,1))^2 + . . .
           (cbs(j)-cc(j,2))^2 + . . .
           (ccs(j)-cc(j,3))^2 ;
end
end

function [dC] = ode_ex8_06(t,C,kpar)
k = kpar(1);
Ka = kpar(2);
Kb = kpar(3);
Kc = kpar(4);
Ca=C(1);
Cb=C(2);
Cc=C(3);
num=k*Ka*Kb*Ca*Cb;
den=(1+Ka*Ca+Kb*Cb+Kc*Cc)^2;
r=num/den;
dC(1)=-r;
dC(2)=-r;
```

```
dC(3)=+r;
dC=dC';
end
```

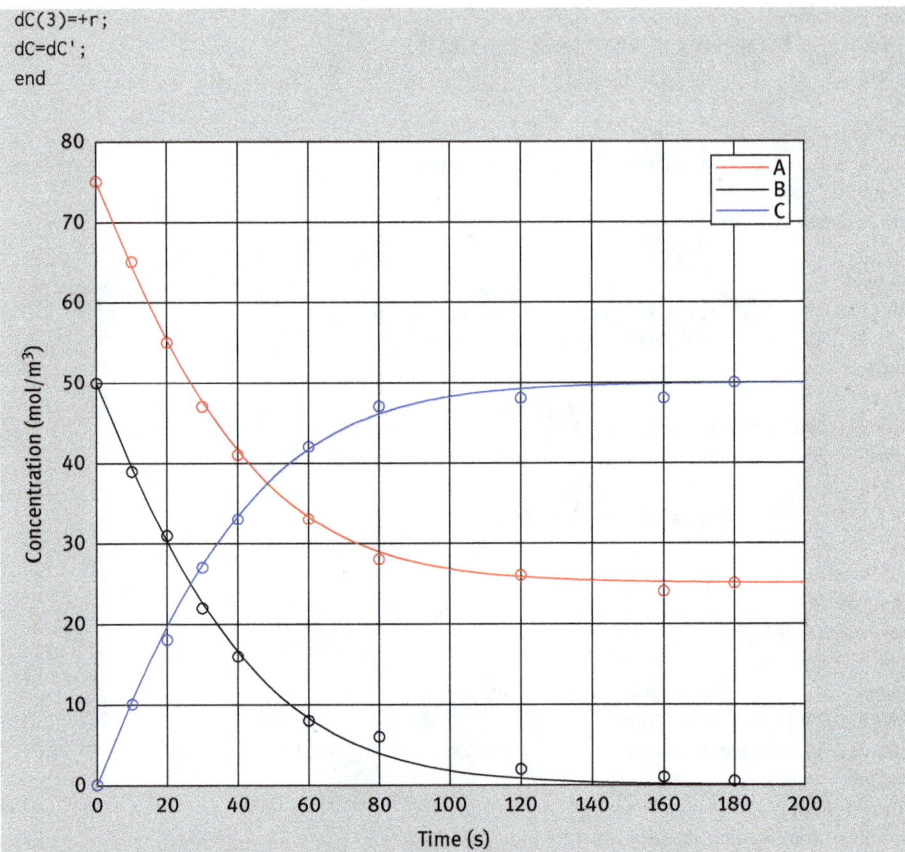

Figure 7.9: Experimental and calculated concentration of A, B and C versus time. Dots are experimental values, and continuous curves represent model prediction.

```
                              Best      Mean    Stall
Iteration   f-count           f(x)        f(x)   Iterations
    0          30         120.2         3656      0
    1          60         120.2     2.626e+04       0
    2          90         120.2     1.499e+04       1
    3         120         37.97       1.6e+04      0
    4         150         25.89         8278     0
    5         180         25.89      1.591e+04        1
    .          .             .          .     .
   66        2010          20.91          346     17
   67        2040          20.91        215.3     18
---------------------------------------------
----------------- Fitted Parameters --------------------
----------------------------------------------------------
```

```
Parameter k 1 = 7.24297e+00
Parameter k 2 = 5.44510e-01
Parameter k 3 = 2.07139e-01
Parameter k 4 = 1.55988e-01
```

List of symbols

f	Function defining the model (algebraic or differential)	[–]
x	Vector of the *N*s experimental observations, independent variables	[*]
$\hat{y}$	Vector of calculated dependent variables (response of the model)	[*]
y_i	Vector of experimental dependent variables	[*]
$\overline{y}$	Average value of experimental *y* data	[*]
O	Objective function	[*]
N_S	Number of experimental observations	[–]
P	Number of adjustable parameters	[–]
w_i	Weight on the *i*th experimental observation	[*]
x_i	Experimental fractional conversion	[–]
T_i	Experimental temperature	[K]
$\vec{x}_i$	Vector of position of the *i*th particle	[*]
x_{ij}	*j*th coordinate of particle *i*	[*]
$\vec{p}_i$	Vector of the best position found by particle *i*	[*]
p_{ij}	*j*th coordinate of the best position of particle *i*	[*]
$\vec{v}_i$	Vector of velocity of particle *i*	[*]
v_{ij}	*j*th component of the velocity of particle *i*	[*]
C	Constant for updating particle swarm velocity	[–]
e_1, e_2	Random numbers	[–]
w_x	Weight related to conversion in objective function	[–]
w_T	Weight related to temperature in objective function	[–]
J	Jacobian matrix	[*]
$J_{i,j}$	Element of Jacobian matrix	[*]
n	Number of experimental observation	[–]
V	Covariance matrix	[*]
V_{ij}	Element of covariance matrix	[*]
S	Standard error	[*]
t_{inv}	Inverse of Student's distribution	[–]
Corr	Correlation matrix	[*]
Corr_{ij}	Element of correlation matrix	[*]
R	Correlation coefficient	[–]
R_{adj}	Adjusted correlation coefficient	[–]
RSS	Residual sum of squares	[*]
TSS	Total sum of squares	[*]
ESS	Estimated sum of squares	[*]
F, F_E	Fisher tests	[–]

F_M	Modified Fisher's test	[–]
f_i	Parameter for normal probability plot	[–]
s_i	Score for the *i*th residual	[–]
z	Variable for standardized normal distribution	[–]

Greek letters

α	Risk for confidence intervals	[–]
β	Vector of p unknown parameters that must be estimated	[*]
ε_i	Residual for *i*th observation	[*]
$\varphi(z)$	Standardized normal distribution	[–]

*Units depend on the variable or the specific system.

References

[1] P. Englezos, N. Kalogerakis. Applied Parameter Estimation for Chemical Engineering. Marcel Dekker Inc.: 2001.

[2] E. B. Nauman. Chemical Reactor Design, Optimization, and Scale-up (2th edition). Wiley: 2008.

[3] A. Bjork. Numerical Methods for Least Squares Problems. Siam: 1996.

[4] A. Tarantola. Inverse Problem Theory and Methods for Model Parameter Estimation. Siam: 2005.

[5] J. A. Snyman. Practical Mathematical Optimization. Springer: 2005.

[6] G. Chavent. Nonlinear Least Squares for Inverse Problems. Springer: 2009.

[7] R. C. Aster, B. Borchers, C. H. Thurber. Parameter Estimation and Inverse Problems. Elsevier: 2005.

[8] T. Back, D.B. Fogel. Handbook of Evolutionary Computation. Michaelewicz ed.: 1997.

[9] K. Marti. Stochastic Optimization Methods. Springer, New York – Berlin: 2005.

[10] A. Zhigljavsky, A. Zilinskas. Stochastic Global Optimization. Springer: 2008.

[11] K. Levenberg. A method for the solution of certain problems in least-squares. Quarterly Applied Math. 1944, 2, 164–168.

[12] D. Marquardt. An algorithm for least-squares estimation of nonlinear parameters. SIAM Journal Applied Math. 1963, 11, 431–441.

[13] J. A. Nelder, R. Mead. A Simplex method for function minimization. The Computer Journal 1965, 7, 308.

[14] J. Kennedy, R. Eberhart. Particle swarm optimization. Proceedings of the IEEE International Conference on Neural Networks. Perth, Australia 1995, 1942–1945.

[15] https://www.itl.nist.gov/div898/handbook/ (last visited 24/06/2020).

[16] V. Russo, O. Ortona, R. Tesser, L. Paduano, M. Di Serio. On the importance to choose the Best Minimization algorithm for the determination of ternary diffusion coefficients by Taylor dispersion method. ACS Omega 2017, 2, 2945–2952.

Chapter 8
Statistical techniques applied to chemical processes

8.1 Introduction

In this section some widely used statistical techniques are described and some solved examples are presented. Two topics are treated: (i) optimization of chemical process and (ii) design of experiment method (DOE).

The first topic deals with specific mathematical formulation of problems concerning processes, and more specifically chemical reactors, in which the minimum or maximum of a function is searched under some constraints. This kind of problem can arise when in a reactor a maximum yield of a particular product must be obtained or when a minimum production cost per unit of mass must be attained. However, optimization studies can be related both to a single process unit, a reactor for example, or to a section or a complete plant.

The second topic treated in this chapter is a low-level statistical modeling technique that can be applied to a very wide range of problems and should be adopted as the basic or preliminary step in any experimental study [1, 2]. In addition, this technique can be used also for industrial application such as formulations or mixtures optimization, in fine tuning of plant conditions and in product quality optimization. In the chemical reaction engineering context, the DOE approach could be used as the first tool to establish the range of optimal conditions in which the reactor can be optimally operated.

8.2 Optimization

As we have already seen in the introduction section, an optimization problem consists, ultimately, of a minimum/maximum search of an objective function of several variables subjected to some constraints. In order to establish a good basis for understanding optimization topic, few definitions must be stated.

- Decision variables or design variables (independent variables). Are the variables that can be specified and from which the objective function depends. Decision variables can be continuous, like pressure and temperature for example, or discrete, like for example the number of batches per day in a discontinuous production plant.
- Objective function. Represents the core of the optimization problem. Is a mathematical function that depends on the decision variables and that should be minimized or maximized. In correspondence to the minimum or maximum of

https://doi.org/10.1515/9783110632927-008

this function, the decision variables are at their optimum. The objective function can be, for example, the sum of squares between experimental and calculated values in a fitting problem (see Chapter 7), a cost function if the economical income must be maximized or the yield of a specific product in a reaction optimization problem.
- Constraints. The constraints are externally imposed limitations on the decision variable or their functions. Can be classified into three distinct categories:
  - Simple bounds on decision variables like $x_{Li} \le x_i \le x_{Ui}$, where x_i is the decision variables that can be varied only in the range $x_{Li} - x_{Ui}$.
  - Equality constraints. Can involve more than one decision variable and can be arranged in a form of a linear system. In a compact form can be expressed as follows:

$$A_{i,j}x_i = b_j \quad i = 1, \ldots, N_v \quad j = 1, \ldots, N_{ec}$$

where N_v is the number of decision variables and N_{ec} is the number of equality constraints. Substantially a rectangular linear system and A_{ij} is a matrix and b_j is a vector. For example, the required stoichiometric oxygen amount in a combustion reactor could be considered as an equality constraint.
  - Inequality constraints. Are limitation imposed as inequalities and can be written in a form very similar to that related to equality constraints. The form is the following:

$$A_{i,j}x_i \le b_j \quad i = 1, \ldots, N_v \quad j = 1, \ldots, N_{ic}$$

Here N_{ic} is the number of inequality constraints. An example of this kind of constraint could be the maximum allowable temperature for avoiding degradation of catalyst or the maximum pressure of gas for safety issues:

$$T_{reactor} \le 400 \text{ K}$$

$$P_{reactor} \le 50 \text{ bar}$$

Strictly speaking, these kinds of simple constraints can be treated as bound constraints. Another example is that of a production line that can use a maximum of 8 t/day of component A for producing X_1 and X_2 amounts of two mixtures, each of them with a specific composition (w_{A1} and w_{A2}). In this case, the inequality constraint can be expressed as follows:

$$w_{A1}X_1 + w_{A2}X_2 \le 8$$

These definitions are particularly useful in the initial phase of the optimization study as they allow to select the proper software tool or code for the solution of the problem. Moreover, the choice of the algorithm is driven by the specific problem characteristics. According to the literature references [3, 4], the classification of optimization problems can be done as follows:

- Linear programming. The objective function is linear with respect to all the decision variables and only linear constraints are present.
- Nonlinear programming. The objective function is intrinsically nonlinear in the decision variables and/or nonlinear equality and noninequality constraints are present. A particular case is when the objective function has a functionality of the second order with respect to some of the decision variables. In this case the problem is defined as quadratic programming.
- Optimization problems that involve both discrete and continuous variables is called mixed integer programming and depending on the linear or nonlinear structure of the objective function the problem is classified as follows:
  - mixed integer linear programming
  - mixed integer nonlinear programming

For a practical approach in problems solution, in MATLAB many functions are present that allow the user to solve essentially all the optimization problems described in the earlier classification.

In the chemical reaction engineering, the majority of the encountered optimization problems can be classified as nonlinear optimization because the objective function can involve the numerical solution of nonlinear equations or differential equations.

Example 8.1 Optimization of a temperature profile in a batch reactor for maximum yield
In a constant volume batch reactor, two simultaneous reactions occur with the following scheme:

$$A + B \rightarrow C$$
$$2B \rightarrow D$$

The problem consists of maximizing the yield of component C by adopting an optimal temperature profile along the batch time instead of a temperature held fixed for all the time. Compare the optimized yield with the one obtained at constant temperature. To build the optimal temperature profile, divide the batch time in 10 intervals and adopt a suitable temperature in each of these intervals.

The kinetic parameters and initial concentrations in the reactor are reported in Table 8.1.

Table 8.1: Kinetic parameters and other constants.

Parameter	Value	Units
Batch time	5,400	s
K_{10}	7.55e08	$m^3/(kmol\ s)$
K_{20}	5.75e11	$m^3/(kmol\ s)$

Table 8.1 (continued)

Parameter	Value	Units
E_{a1}	79,000	kJ/kmol
E^{a2}	98,000	kJ/kmol
Vr	1	m^3
T_L lower limit	313.15	K
T_U upper limit	393.15	K
T fix (reference case)	350.15	K

For a complete definition of the problem, the expressions of the liquid-phase reaction rates are necessary, together with the values of the corresponding kinetic constants as functions of temperature:

$$r_1 = k_1 C_A C_B \quad k_1 = k_{10} \exp(-E_{a1}/RT)$$
$$r_2 = k_2 C_B^2 \quad k_2 = k_{20} \exp(-E_{a2}/RT)$$

Develop a Matlab code that solves the differential material balance equations draw the following plots as subplots array:

1) Concentration profiles of all the components along the reaction batch time at $T_{fix} = 350.13$ K.
2) Yield of C at T fix
3) Yield of C at fixed temperature in the range T_{min}–T_{max}
4) Concentration profiles of all the components along the reaction batch time with variable temperature
5) Optimized yield
6) Optimized temperature profile

In a compact form, the material balance for the batch reactor can be written as follows:

$$\frac{dC_i}{dt} = \sum_{k=1}^{Nr} v_{i,k} r_k \quad i = \text{A, B, C, D} \quad N_r = 2 \quad \text{reactions}$$

The yield of C is defined as follows:

$$y_C = \frac{C_{C,t=5,400s}}{C_{B,t=0}}$$

The optimization problem can be defined with the following formal notation:

$$\text{Max} \quad y_C$$
$$\text{Adjustable parameters: } T_i \qquad i = 1, .., 10$$
$$\text{Constraints:} \qquad T_L \le T_i \le T_U$$

Matlab code for the solution of this example and results are reported as follows:

```
% example 8.1
clc,clear
close all
%% parameters
k10 = 7.55e8;           % pre-exp factor 1 (m3/(kmol/s))
k20 = 5.75e11;          % pre-exp factor 1 (m3/(kmol/s))
Ea1 = 79000;            % activation energy reaction 1 (kj/kmol)
Ea2 = 98000;            % activation energy reaction 2 (kj/kmol)

T   = 350.15;           % fixed temperature (K)
Vol  = 1;               % reactor volume (m3)
Batch_time = 5400;      % batch time (s)

%% simulation
tspan=0:0.01:Batch_time;
Ca0=1;
Cb0=1;
Cc0=0;
Cd0=0;
C0=[Ca0 Cb0 Cc0 Cd0];
jopt=0;
ode = @(t,C)ode_ex8_01(t,C,k10,k20,Ea1,Ea2,T,1,jopt);
[t,C]=ode15s(ode,tspan,C0);

%% Parametric study at constant temperature
Tx=313.15:2:393.15;
jopt=0;
tic
parfor j=1:length(Tx)
   T=Tx(j);
   ode = @(t,C)ode_ex8_01(t,C,k10,k20,Ea1,Ea2,T,1,jopt);
   [tx,Cx]=ode15s(ode,tspan,C0);
   Ycx=Cx(:,3)/Cb0;
   YYx(j)=Ycx(end);
end
toc

%% optimization
jopt=1;
NT=10;
Tmin=313.15;
```

```
Tmax=393.15;
Tj=linspace(Tmin+1,Tmax-1,NT);
TL=ones(NT,1)*Tmin;
TU=ones(NT,1)*Tmax;
options = optimoptions('fmincon','Display','iter');
obj = @(Tj)objfun_ex8_01(Tj,Batch_time,NT,jopt,k10,k20,Ea1,Ea2);
[Topt,fval]=fmincon(obj,Tj,[],[],[],[],TL,TU,[],options)
T_step=Batch_time/NT;
v=zeros(NT,3);
for j=1:NT
   v(j,2)=T_step*j;
   v(j,3)=Topt(j);
end
for j=2:NT
   v(j,1)=v(j-1,2);
end
tspan=0:0.01:Batch_time;
Ca0=1;
Cb0=1;
Cc0=0;
Cd0=0;
C0=[Ca0 Cb0 Cc0 Cd0];
ode = @(t,C)ode_ex8_01(t,C,k10,k20,Ea1,Ea2,1,v,jopt);
[t,Copt]=ode15s(ode,tspan,C0);

%% plots
subplot(2,3,1)
plot(t,C)
grid
xlabel('Time (s)')
ylabel('Concentration (mol/m3)')
legend('A','B','C','D')
title('1')

subplot(2,3,2)
Yc=C(:,3)/Cb0;
plot(t,Yc)
grid
xlabel('Time (s)')
ylabel('Yield of C (-)')
title('2')

subplot(2,3,3)
plot(Tx,YYx,'ro-')
```

```
grid
xlabel('T constant (K)')
ylabel('Yield of C (-)')
title('3')

subplot(2,3,4)
plot(t,Copt)
grid
xlabel('Time (s)')
ylabel('Concentration (mol/m3)')
legend('A','B','C','D')
title('4')

subplot(2,3,5)
Yc_opt=Copt(:,3)/Cb0;
plot(t,Yc_opt)
grid
xlabel('Time (s)')
ylabel('Optimized yield of C (-)')
title('5')

subplot(2,3,6)
plot(v(:,1),v(:,3),'db-')
grid
xlabel('Time (s)')
ylabel('Optimized temperature (K)')
title('6')

function [f] = objfun_ex8_01(Tj,Batch_time,NT,jopt,k10,k20,Ea1,Ea2)
T_step=Batch_time/NT;
v=zeros(NT,3);
for j=1:NT
   v(j,2)=T_step*j;
   v(j,3)=Tj(j);
end
for j=2:NT
   v(j,1)=v(j-1,2);
end

%% integration
tspan=0:0.1:Batch_time;
Ca0=1;
```

```
Cb0=1;
Cc0=0;
Cd0=0;
C0=[Ca0 Cb0 Cc0 Cd0];
ode = @(t,C)ode_ex8_01(t,C,k10,k20,Ea1,Ea2,1,v,jopt);
[t,C]=ode15s(ode,tspan,C0);
Yc=C(:,3)/Cb0;
f=-Yc(end);
end

function [dC] = ode_ex8_01(t,C,k10,k20,Ea1,Ea2,T,v,jopt)
if jopt==0
   Tk=T;
end
if jopt==1
   Tk=Tstep(t,v);

end
Ca = C(1);
Cb = C(2);
Cc = C(3);
Cd = C(4);
R = 8.31446;
k1 = k10*exp(-Ea1/(R*Tk));
k2 = k20*exp(-Ea2/(R*Tk));
r1 = k1*Ca*Cb;
r2 = k2*Cb^2;
dC(1) = -r1;
dC(2) = -r1-2*r2;
dC(3) = +r1;
dC(4) = +r2;
dC=dC';
end

function [Tt] = Tstep(t,v)
%% Temperature profile vs time
% t   time      (s)
% v   tabel of values (s  s  K)
%-----------------------------------------------
Tt=0;
[rv,~]=size(v);  % number of rows in table v
for j=1:rv
  if t>=v(j,1) && t<v(j,2)
```

```
    Tt=v(j,3);
    break
    end
  end

  end
```

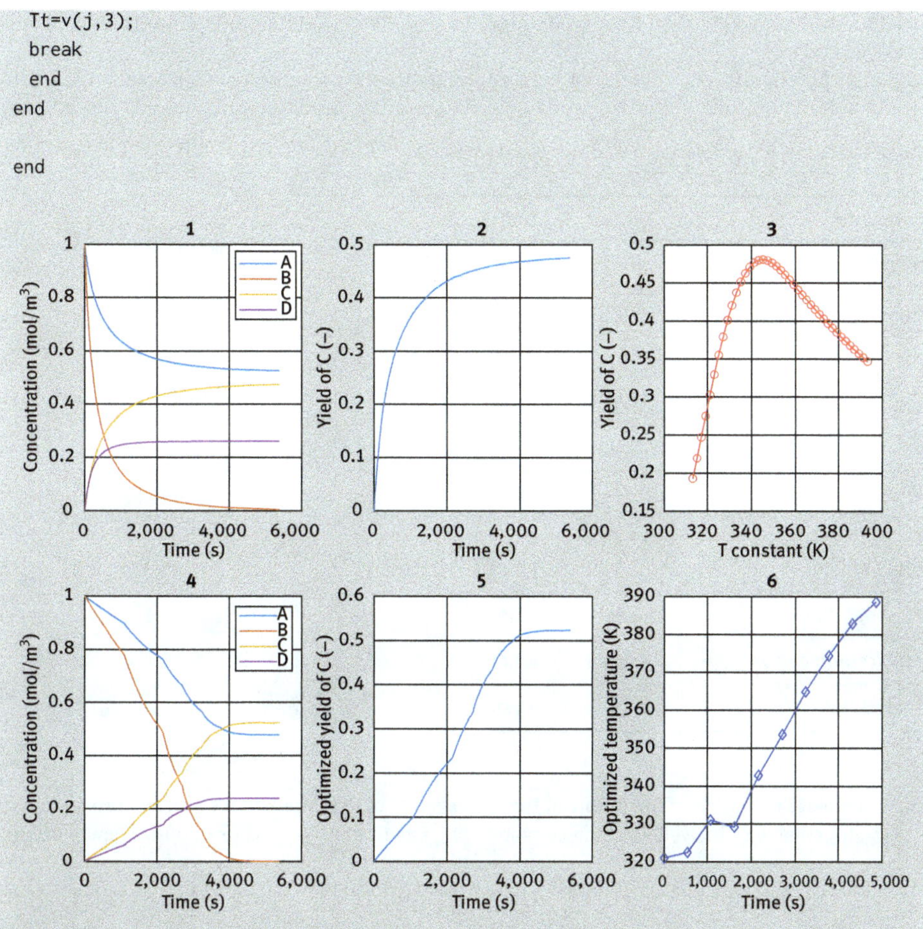

Figure 8.1: From top to left: (1) concentration profile at $T = 350$ K; (2) yield of C at $T = 350$ K; (3) yield of C as a function of a fixed temperature; (4) concentration profiles for optimized temperature profile; (5) optimized yield profile; (6) optimized temperature profile.

Example 8.2 Optimization of a feed profile in a fed-batch reactor for maximum yield
In a fed-batch reactor two simultaneous reactions occurs with the following scheme:

$$A + B \longrightarrow C$$

$$2B \longrightarrow D$$

The problem consists in maximizing the yield of component C by adopting an optimal feed profile by simultaneously keeping fixed temperature value for all the time. Compare the optimized yield with the one obtained by keeping fixed the feed flowrate. To build the optimal feed profile, divide the operation time in 10 intervals and adopt a suitable feed flowrate in each of these intervals.

Kinetic parameters, constraints and initial concentrations in the reactor are reported in Table 8.2.

Table 8.2: Kinetic parameters and other constants.

Parameter	Value	Units
Batch time	3,600	s
K_{10}	7.55e08	$m^3/(kmol\ s)$
K_{20}	5.75e11	$m^3/(kmol\ s)$
E_{a1}	79,000	kJ/kmol
E^{a2}	98,000	kJ/kmol
Vr	1	m^3
T F_L lower limit	393.15 0	K kmol/s
F_U upper limit	5.0e-04	kmol/s
F fix (reference case)	2.778e-04	kmol/s
Total B fed	1	kmol

For a complete definition of the problem, the expressions of the liquid-phase reaction rates are necessary, together with the values of the corresponding kinetic constants as functions of temperature:

$$r_1 = k_1 C_A C_B \quad k_1 = k_{10} \exp(-E_{a1}/RT)$$

$$r_2 = k_2 C_B^2 \quad k_2 = k_{20} \exp(-E_{a2}/RT)$$

Develop a Matlab code that solves the differential material balance equations draw the following plots as subplots array:

1) Concentration profiles of all the components along the reaction batch time at $F\text{fix} = 2.778\text{e-}4$ kmol/s
2) Yield of C for fixed feed and for optimized feed profile
3) Concentration profiles of all the components along the reaction batch time with variable feed profile
4) Optimized feed profile

In a compact form, the material balance for the fed-batch reactor and for components that are not fed can be written as

$$\frac{dn_i}{dt} = V_R \sum_{k=1}^{Nr} v_{i,k} r_k \quad i = \text{A, C, D} \quad N_r = 2 \quad \text{reactions}$$

While for B the material balance is as follows:

$$\frac{dn_B}{dt} = V_R \sum_{k=1}^{Nr} v_{B,k} r_k + F_B \quad N_r = 2 \quad \text{reactions}$$

The yield of C is defined as follows:

$$y_C = \frac{n_{C,t=3,600\,s}}{n_{B,t=3,600}}$$

The optimization problem can be defined with the following formal notation:

$$\text{Max} \quad y_C$$

$$\text{Adjustable parameters: } F_i \qquad i = 1, \ldots, 10$$

$$\text{Constraints:} \qquad F_L \le F_i \le F_U$$

$$\int_{t=0}^{t=3,600} F_B(t)dt = 1$$

Matlab code for the solution of this example and results are reported as follows:

```
% example 8.2
clc,clear
close all
%% parameters
k10 = 7.55e8;
k20 = 5.75e11;
Ea1 = 79000;
Ea2 = 98000;
T   = 393.15;
Vol = 1;
Batch_time = 3600;
%% simulation
tspan=0:0.01:Batch_time;
Ca0=1;
Cb0=0;
Cc0=0;
Cd0=0;
na0=Ca0*Vol;
nb0=Cb0*Vol;
nc0=Cc0*Vol;
```

```
nd0=Cd0*Vol;
C0=[Ca0 Cb0 Cc0 Cd0];
n0=[na0 nb0 nc0 nd0 nb0];
jopt=0;
Fb=2.7778e-4;   % kmol/s
ode = @(t,n)ode_ex8_02(t,n,k10,k20,Ea1,Ea2,T,1,Fb,Vol,jopt);
[t,n]=ode15s(ode,tspan,n0);
C=n/Vol;
Fbs=n(:,5);
Yc=C(:,3)./Fbs(end);
%% optimization
jopt=1;
NF=10;
Fmin=0;
Fmax=5e-4;
epsF=1e-7;
F0=[38 37 36 34 33 31 27 25 21 5]*1e-5;
Fj=F0;
FL=ones(NF,1)*Fmin;

FU=ones(NF,1)*Fmax;
options = optimoptions('fmincon','Display','iter','Algorithm','interior-
point',...
        'UseParallel',true);
obj = @(Fj)objfun_ex8_02(Fj,Batch_time,NF,jopt,k10,k20,Ea1,Ea2,T,Vol);
nlc = @(Fj)nonlcon_ex8_02(Fj,Batch_time,NF,jopt,k10,k20,Ea1,Ea2,T,Vol);

[Fopt,fval]=fmincon(obj,Fj,[],[],[],[],FL,FU,nlc,options)
F_step=Batch_time/NF;
v=zeros(NF,3);
for j=1:NF
   v(j,2)=F_step*j;
   v(j,3)=Fopt(j);
end
for j=2:NF
   v(j,1)=v(j-1,2);
end
tspan=0:0.01:Batch_time;
Ca0=1;
Cb0=0;
Cc0=0;
Cd0=0;
```

```
na0=Ca0*Vol;
nb0=Cb0*Vol;
nc0=Cc0*Vol;
nd0=Cd0*Vol;

C0=[Ca0 Cb0 Cc0 Cd0];
n0=[na0 nb0 nc0 nd0 nb0];

ode = @(t,n)ode_ex8_02(t,n,k10,k20,Ea1,Ea2,T,v,1,Vol,jopt);
[t,nopt]=ode15s(ode,tspan,n0);
Copt=nopt/Vol;
nopt(end,5)
nbalim=nopt(:,5);
Yc_opt=nopt(:,3)./nbalim(end);

%% plots
subplot(2,2,1)
plot(t,C(:,1:4))
grid
xlabel('Time (s)')
ylabel('Concentration (mol/m3)')


legend('A','B','C','D')
title('1')

subplot(2,2,2)
plot(t,Yc,t,Yc_opt)
grid
xlabel('Time (s)')
ylabel('Yield of C (-)')
legend('Yield by simul.','Yield by optim.')
title('2')

subplot(2,2,3)
plot(t,Copt(:,1:4))
grid
xlabel('Time (s)')
ylabel('Concentration (mol/m3) (optim.)')
legend('A','B','C','D')
title('3')
```

```
subplot(2,2,4)
for j=1:length(t)
   ft(j)=Fstep(t(j),v);
end
plot(t,ft)
grid
xlabel('Time (s)')
ylabel('Optimized B feed rate (kmol/s)')
title('4')

function [f] = objfun_ex8_02(Fj,Batch_time,NF,jopt,k10,k20,Ea1,Ea2,T,Vol)
F_step=Batch_time/NF;
v=zeros(NF,3);
for j=1:NF
   v(j,2)=F_step*j;
   v(j,3)=Fj(j);
end
for j=2:NF
   v(j,1)=v(j-1,2);
end
%% integration
tspan=0:0.1:Batch_time;
Ca0=1;


Cb0=0;
Cc0=0;
Cd0=0;
na0=Ca0*Vol;
nb0=Cb0*Vol;
nc0=Cc0*Vol;
nd0=Cd0*Vol;
C0=[Ca0 Cb0 Cc0 Cd0];
n0=[na0 nb0 nc0 nd0 nb0];
ode = @(t,n)ode_ex8_02(t,n,k10,k20,Ea1,Ea2,T,v,1,Vol,jopt);
[t,n]=ode15s(ode,tspan,n0);
nbalim=n(:,5);
Yc=n(:,3)./nbalim(end);
f=-Yc(end);
end

function [c,ceq] = nonlcon(Fj,Batch_time,NF,jopt,k10,k20,Ea1,Ea2,T,Vol)
F_step=Batch_time/NF;
v=zeros(NF,3);
```

```
for j=1:NF
   v(j,2)=F_step*j;
   v(j,3)=Fj(j);
end
for j=2:NF
   v(j,1)=v(j-1,2);
end

%% integration
tspan=0:0.1:Batch_time;
Ca0=1;
Cb0=0;
Cc0=0;
Cd0=0;
na0=Ca0*Vol;
nb0=Cb0*Vol;
nc0=Cc0*Vol;
nd0=Cd0*Vol;
C0=[Ca0 Cb0 Cc0 Cd0];
n0=[na0 nb0 nc0 nd0 nb0];
ode = @(t,n)ode_ex8_02(t,n,k10,k20,Ea1,Ea2,T,v,1,Vol,jopt);
[t,n]=ode15s(ode,tspan,n0);

nbalim=n(:,5);
ceq   = nbalim(end)-1;
c     = [];
end

function [dn] = ode_ex8_02(t,n,k10,k20,Ea1,Ea2,T,v,Fb,Vol,jopt)
if jopt==0
   Fbk=Fb;
end
if jopt==1
   Fbk=Fstep(t,v);
end
C=n/Vol;
Ca = C(1);
Cb = C(2);
Cc = C(3);
Cd = C(4);
R = 8.31446;
k1 = k10*exp(-Ea1/(R*T));
k2 = k20*exp(-Ea2/(R*T));
r1 = k1*Ca*Cb;
r2 = k2*Cb^2;
dn(1) = Vol*(-r1);
dn(2) = Fbk + Vol*(-r1 - 2*r2);
dn(3) = Vol*(+r1);
dn(4) = Vol*(+r2);
```

```
dn(5) = Fbk;
dn=dn';
end

function [Ft] = Fstep(t,v)
Ft=0;
[rv,~]=size(v);          % number of rows in table v
for j=1:rv
  if t>=v(j,1) && t<v(j,2)
    Ft=v(j,3);
    break
  end
end
end
```

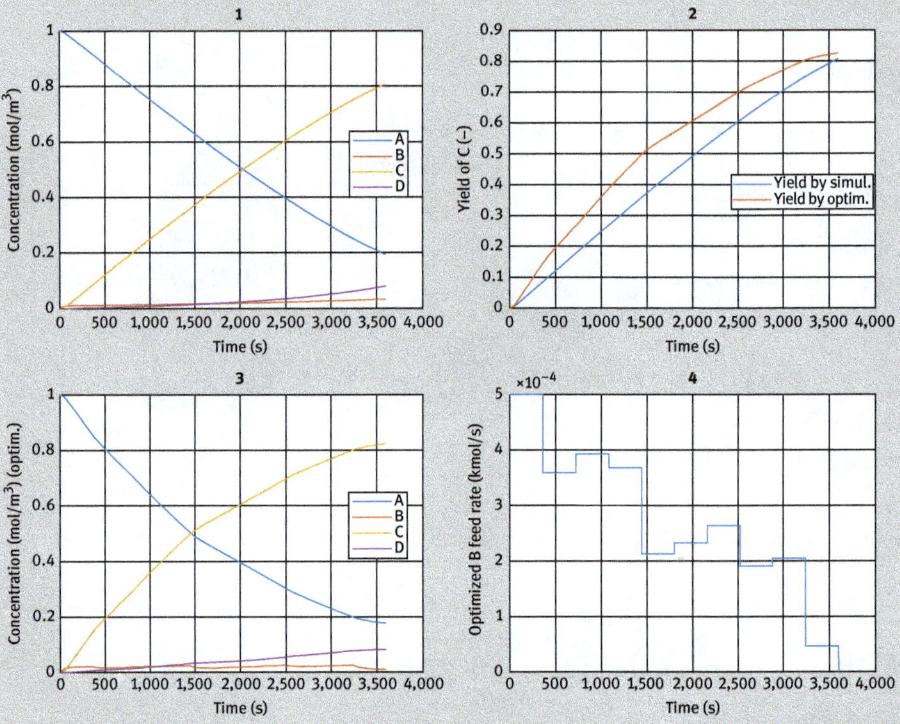

Figure 8.2: From top to left: (1) concentration profile at a fixed feed of 2.778e-04 kmol/s; (2) yield of C for a fixed feed and for optimized feed profile; (3) concentration profiles for optimized feed profile; (4) optimized feed profile.

Example 8.3 Optimization of a thermal cracking unit.

An industrial unit dedicated to thermal cracking of different feedstocks must be optimized in terms of economical incomes. The scheme of this unit is reported in Figure 8.3 from which it is possible to see the different feeds, different products and the recycle strategy.

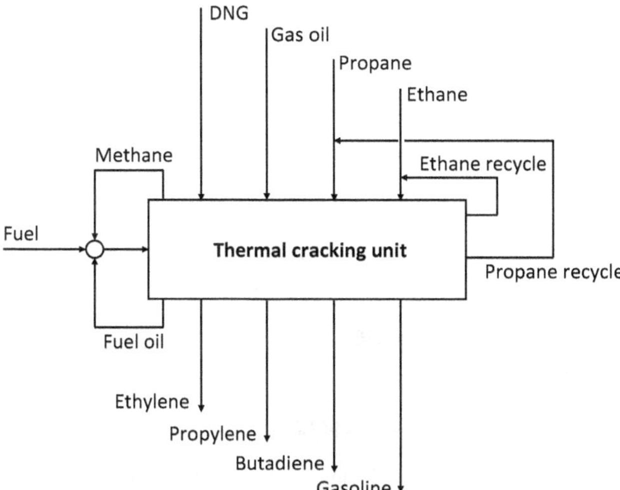

Figure 8.3: Scheme of thermal cracking unit.

This unit has been studied in the literature [4] and the economic function, that should be maximized, is the following:

$$f(x_1, x_2, x_3, x_4, x_5, x_6, x_7) = 2.84x_1 - 0.22x_2 - 3.33x_3 + 1.09x_4 + 9.39x_5 + 9.51x_6$$

The meaning of the design variable is defined in the following list:

- x_1 fresh ethane feed (lb/h)
- x_2 fresh propane feed (lb/h)
- x_3 gas oil feed (lb/h)
- x_4 debutanized natural gas (DNG) feed (lb/h)
- x_5 ethane recycled (lb/h)
- x_6 propane recycled (lb/h)
- x_7 fuel added (lb/h)

Note that even if formally this function depends on seven design variables, only the first six of them appear really in the function. The maximization of the function that represent the economical advantage of the unit must be performed under some constraints expressed in terms of inequalities and equalities.

Case 1

Only one inequality constraint is present that corresponds to the maximum capacity of the unit to treat feedstock amount: 200.000 lb/h. This constraint can be expressed as follows:

$$1.1x_1 + 0.9x_2 + 0.9x_3 + 1.0x_4 + 1.1x_5 + 0.9x_6 \leq 200,000$$

The remaining constraints for the optimization problem are defined as equalities and in more detail as
Total ethylene production = 50,000 lb/h

$$0.50x_1 + 0.35x_2 + 0.20x_3 + 0.25x_4 + 0.50x_5 + 0.35x_6 = 50000$$

Total propylene production = 20,000 lb/h

$$0.01x_1 + 0.15x_2 + 0.15x_3 + 0.18x_4 + 0.01x_5 + 0.15x_6 = 20000$$

Ethane recycle:

$$0.40x_1 + 0.06x_2 + 0.04x_3 + 0.05x_4 - 0.60x_5 + 0.06x_6 = 0$$

Propane recycle:

$$0.10x_2 + 0.01x_3 + 0.01x_4 - 0.90x_6 = 0$$

Energy constraint = 20e6

$$-6857.6x_1 + 364x_2 + 2032x_3 - 1145x_4 - 6857.6x_5 + 364x_6 + 21520x_7 = 20 \cdot 10^6$$

The problem can be solved in MATLAB through the use of "linprog" function where both objective function and constraints can be introduced in the code as vector and matrix coefficients.

Case 2

The case 2 is practically the same as case 1 with the exception that two linear constraints (ethylene and propylene production were treated as inequalities instead of equality constraints:
Total ethylene production ≤ 50,000 lb/h

$$0.50x_1 + 0.35x_2 + 0.20x_3 + 0.25x_4 + 0.50x_5 + 0.35x_6 \leq 50,000$$

Total propylene production ≤ 20,000 lb/h

$$0.01x_1 + 0.15x_2 + 0.15x_3 + 0.18x_4 + 0.01x_5 + 0.15x_6 \leq 20,000$$

The output of the Matlab code for both cases is the following:
Optimal solution found.

Results - Case 1

Cost objective function:	298,582.18
x(1) fresh ethane	21,767.87
x(2) fresh propane	1.00
x(3) gas oil feed	1.00
x(4) DNG feed	107,592.80
x(5) ethane recycle	23,597.71
x(6) propane recycle	1195.60
x(7) fuel added	21,089.92

Constraints

Inequality constraint, max capacity : 158,572.78 ≤ 200,000

Inequality constraint, max capacity	: 158,572.78 ≤ 200,000	
Ethylene limitation	: 50,000.00 =	50,000
Propylene limitation	: 20,000.00 =	20,000
Ethane recycle	: -0.00 =	0
Propylene recycle	: -0.00 =	0
Energy constraint	: 20,000,000.00 =	20e6

Optimal solution found.

Results - Case 2

Cost objective function:	369,541.82
x(1) fresh ethane	59,994.67
x(2) fresh propane	8.80
x(3) gas oil feed	1.00
x(4) DNG feed	1.00
x(5) ethane recycle	39,997.57
x(6) propane recycle	1.00
x(7) fuel added	32,792.86

Constraints

Inequality constraint, max capacity:	110,002.18 ≤ 200,000
Inequality constraint, ethylene limitation:	50,000.00 ≤ 50,000
Inequality constraint, propylene limitation:	1001.72 ≤ 20,000

Ethane recycle	:	−0.00 = 0
Propylene recycle	:	−0.00 = 0
Energy constraint	:	20,000,000.00 = 20e6

>>

Matlab code for the solution of this example and results are reported as follows:

```
% example 8.3
clc,clear
%% CASE 1
%% objective function
objf=-1*[2.84 -0.22 -3.33 +1.09 +9.39 +9.51 0];
```

```
%% inequality constraints
A=[ 1.1  0.9  0.9  1.0  1.1  0.9  0.0];
b=200000;
%% equality constraints
Aeq = [ 0.50  0.35  0.20  0.25  0.50  0.35    0.00 ;
    0.01   0.15  0.15  0.18   0.01  0.15  0.00 ;
    0.40   0.06  0.04  0.05  -0.60  0.06  0.00 ;
    0.0    0.10  0.01  0.01   0.00 -0.90  0.00
   -6857.6  364  2032  -1145    -6857.6  364  21520 ];
beq=[ 50000 ;
   20000 ;
   0    ;
   0    ;
   20e6 ];
%% optimization by linear programmiing
x0=ones(1,7);
[x,fval] = linprog(objf,A,b,Aeq,beq,x0);
fval=-fval-5.38*x(7);

ineq1=A*x;
c=Aeq*x;

%% output
disp(' Results - Case 1')
disp('-------------------------------')
fprintf('  Cost objective function :%10.2f\\n',fval)
disp(' ')
fprintf(' x(1) fresh ethane       %10.2f \n',x(1))
fprintf(' x(2) fresh propane      %10.2f \n',x(2))
fprintf(' x(3) gas oil feed       %10.2f \n',x(3))
fprintf(' x(4) DNG feed           %10.2f \n',x(4))
fprintf(' x(5) ethane recycle     %10.2f \n',x(5))
fprintf(' x(6) propane recycle    %10.2f \n',x(6))
fprintf(' x(7) fuel added         %10.2f \n',x(7))
disp(' ')
disp(' Constraints ')
disp('-------------------------------')
fprintf(' Inequality constraint, max capacity  : %10.2f     <= 200000 \n',ineq1)
disp(' ')
fprintf(' Ethylene limitation    : %12.2f=50000  \n',c(1))
fprintf(' Propylene limitation   : %12.2f=20000  \n',c(2))
fprintf(' Ethane recycle         : %12.2f=0  \n',c(3))
```

```
fprintf(' Propylene recycle     : %12.2f=0  \n',c(4))
fprintf(' Energy constraint     : %12.2f=20e6  \n',c(5))

%% CASE 2

%% inequality constraints
A2=[ 1.10   0.90   0.90   1.00   1.10   0.90   0.00        ;
   0.5    0.35   0.20   0.25   0.50   0.35   0.00          ;
   0.01   0.15   0.15   0.18   0.01   0.15   0.00          ];

b2=[200000;
   50000;
   20000];

%% equality constraints
Aeq2 = [0.40   0.06   0.04   0.05   -0.60   0.06   0.00;
     0.00   0.10   0.01   0.01   0.00   -0.90   0.00;
     -6857.6   364   2032   -1145   -6857.6   364   21520 ];

beq2=[ 0      ;
    0     ;
    20e6   ];

%% optimization by linear programmiing
x0=ones(1,7);
[x,fval] = linprog(objf,A2,b2,Aeq2,beq2,x0);
fval=-fval-5.38*x(7);

ineq2=A2*x;
c2=Aeq2*x;

%% output
disp(' Results - Case 2')
disp('-------------------------------')
fprintf('Cost objective function : %10.2f\n',fval)
disp(' ')
fprintf(' x(1) fresh ethane       %10.2f \n',x(1))
fprintf(' x(2) fresh propane      %10.2f \n',x(2))
fprintf(' x(3) gas oil feed       %10.2f \n',x(3))
fprintf(' x(4) DNG feed           %10.2f \n',x(4))
fprintf(' x(5) ethane recycle     %10.2f \n',x(5))
fprintf(' x(6) propane recycle    %10.2f \n',x(6))
fprintf(' x(7) fuel added         %10.2f  \n',x(7))
```

```
disp(' ')
disp(' Constraints ')
disp('--------------------------------')
fprintf(' Inequality constraint, max capacity:   %10.2f <= 200000 \n',ineq2(1))
fprintf(' Inequality constraint, ethylene limitation: %10.2f <=50000 \n',ineq2(2))
fprintf(' nequality constraint, propylene limitation: %10.2 <=20000 \n',ineq2(3))
disp(' ')
fprintf(' Ethane recycle:       %12.2f  =   0 \n',c2(1))
fprintf(' Propylene recycle:    %12.2f  =   0 \n',c2(2))
fprintf(' Energy constraint:    %12.2f  =   20e6 \n',c2(3))
```

Example 8.4 Optimization of ammonia synthesis reactor

A reactor for the autothermal synthesis of ammonia is operated in a way that the feed is preheated by exchanging heat in the reaction zone. The scheme of the reactor is reported in Figure 8.4:

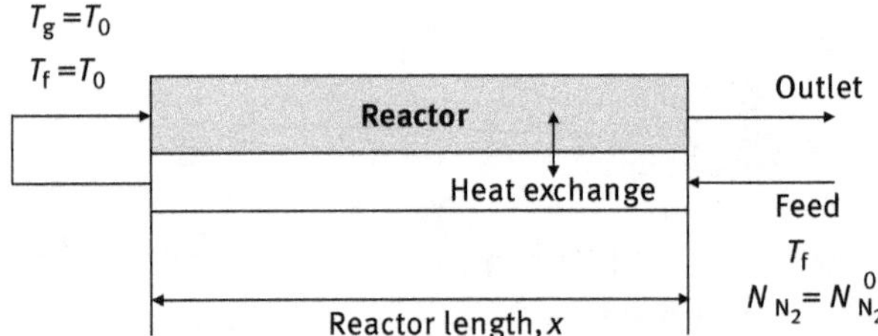

Figure 8.4: Scheme of the fixed bed reactor for ammonia synthesis.

The heat evolved by the reaction:

$$N_2 + 3NH_2 \longleftrightarrow 2NH_3$$

correspond to −26.6 kcal/mol of N_2 converted and is used to preheat the feed sent to the reactor to a temperature at the inlet of the catalytic bed. According to literature studies [5, 6], the objective function suited for the optimization of this reactor is a function of four variables as

$$F_{obj} = f(x, N_{N_2}, T_f, T_g)$$

In more details, the objective function is represented by the economical annual income measured in $/year defined by the relation:

$$F_{obj} = 1.33565e7 - 1.70843e4 + 704.09(T_g - T_0) - 699.27(T_f - T_0) - [3.45663e7 + 1.98865e9x]^{0.5}$$

The model of the reactor is constituted by three ordinary differential equations that represent (i) energy balance on gas feed, (ii) energy balance on the gas in the reactor and (iii) mass balance on nitrogen in the catalytic bed. These equations have the reactor length as independent variable and are

$$\frac{dT_f}{dx} = -\frac{US_1}{WC_{pf}}(T_g - T_f)$$

$$\frac{dT_g}{dx} = -\frac{US_1}{WC_{pg}}(T_g - T_f) + \frac{(-\Delta H)S_2}{WC_{pg}}\left(-\frac{dN_{N_2}}{dx}\right)$$

$$\frac{dN_{N_2}}{dx} = -f_a\left[k_1\frac{P_{N_2}P_{H_2}^{1.5}}{P_{NH_3}} - k\frac{P_{NH_3}}{P_{H_2}^{1.5}}\right]$$

where

$$k_1 = (1.78954E4)e^{-20800/(RT_g)}$$

$$k_2 = (2.57140E16)e^{-47400/(RT_g)}$$

The initial conditions for the dependent variables of the differential equations are

$$T_f = T_0 \quad \text{at} \quad x = 0$$

$$T_g = T_0 \quad \text{at} \quad x = 0$$

$$N_{N_2} = N_{N_2}^0 \quad \text{at} \quad x = 0$$

The optimization problem is defined, for the maximization of objective function F, as follows:

$$\begin{cases} \max F \text{ subjected to contrains:} \\ 400 \le T_0 \le 800\text{K} \\ 0 \le x \le 10\text{m} \end{cases}$$

While $N_{N_2}^0$ is fixed at 701.2 kmol/(h m^2). Other parameters and constants necessary for the solution of the problem are summarized in Table 8.3:

Table 8.3: Parameters and other constants.

Parameter	Value	Units
C_{pf}	0.707	kcal/(kg K)
C_{pg}	0.719	kcal/(kg K)
ΔH	−26,600	kcal/kmol$_{N2}$
R	1.987	kcal/(kmol K)
U	500	kcal/(m$_2$ h K)
f_a	1	–
W	26,400	kg/h
S_1	10	m
S_2	0.78	m^2
Ptot	286	atm

For a complete definition of the problem, the expressions of the partial pressures of nitrogen, hydrogen and ammonia are needed:

$$P_{N_2} = P_{tot} N_{N_2} / (2.598 N^0_{N_2} + 2N_{N_2})$$

$$P_{H_2} = 3P_{N_2}$$

$$P_{NH_3} = P_{tot}(2.23 N^0_{N_2} - 2N_{N_2}) / (2.598 N^0_{N_2} + 2N_{N_2})$$

Develop a Matlab code that solves the optimization problem by finding optimal values of the two decision variables: reactor length x and reactor inlet temperature T_0. Draw the following three plots as subplots array:

1) Feed and reactor temperatures along the reactor length.
2) Nitrogen flow along reactor length.
3) Surface of the objective function as a function of the two decision variables.

Matlab code for the solution of this example and results are reported as follows:

```
% example 8.4
clc,clear
%% data and parameters
  Cpf  = 0.707;
  Cpg  = 0.719;
  fc   = 1;
  DH   = -26600;
  Ptot = 286;
  R    = 1.987;
  S1   = 10;
  S2   = 0.78;
  U    = 500;
  W    = 26400;
NN20   = 701.2;

%% optimization
  LB = [ 0400];
  UB = [10800];
  options = optimoptions('ga','PlotFcn',@gaplotbestf,'Display',. . .
        'Iter','UseParallel',true);
  obj = @(k)objfun_ex8_04(k,Cpf,Cpg,fc,DH,Ptot,R,S1,S2,U,W,NN20);
  k=ga(obj,2,[],[],[],[],LB,UB,[],options)

%% simulation
  xr =k(1);
  T0 =k(2);
  xspan=0:0.001:xr;
```

```
  Tf0 = T0;
  Tg0 = T0;
  y0 =[Tf0 Tg0 NN20];

  opts = odeset('RelTol',1e-6,'AbsTol',1e-8);
  ode = @(x,y)ode_ex8_04(x,y,Cpf,Cpg,fc,DH,Ptot,R,S1,S2,U,W,NN20);
  [x,y] = ode15s(ode,xspan,y0,opts);
  Tf = y(:,1);
  Tg = y(:,2);
  NN2 = y(:,3);

%% surface
  N =50;
  xi = linspace(5,10,N);
  yi = linspace(680,720,N);
  [X,Y] = meshgrid(xi,yi);
  for j=1:N
    parfor j1=1:N
       kk=[X(j,j1) Y(j,j1)];
       Z(j,j1) = -objfun_ex8_ 04(kk,Cpf,Cpg,fc,DH,Ptot,R,S1,S2,U,W,NN20);
     end
  end

%% plots
  subplot(1,3,1)
  plot(x,y(:,1:2))
  grid
  legend('Tf','Tg')
  xlabel('Reactor length (m)')
  ylabel('Temperatures (K)')

  subplot(1,3,2)
  plot(x,y(:,3))
  grid
  xlabel('Reactor length (m)')
  ylabel('N2 flow (kmol/(h m2))')

  subplot(1,3,3)
  surfc(X,Y,Z)
  colormap 'jet'
  view(-26,43)
  xlabel('Reactor length (m)')
  ylabel('Inlet temperature (K)')
  zlabel('Objective function ($/y)')
  title('ps\_example(x)')
  colorbar
```

```
function [g] = objfun_ex8_04(k,Cpf,Cpg,fc,DH,Ptot,R,S1,S2,U,W,NN20)
  xr =k(1);
  T0 =k(2);
  Tf0 = T0;
  Tg0 = T0;
  xspan =0:0.001:xr;
  y0 =[Tf0 Tg0 NN20];
  opts = odeset('RelTol',1e-6,'AbsTol',1e-8);
  ode =@(x,y)ode_ex8_04(x,y,Cpf,Cpg,fc,DH,Ptot,R,S1,S2,U,W,NN20);
  [x,y] =ode15s(ode,xspan,y0,opts);
  Tfe = y(end,1);
  Tge = y(end,2);
  NN2e = y(end,3);
  xe = x(end);
  f = 1.33563e7 - 1.70843e4*NN2e + 704.09*(Tge-T0) . . .
  -699.27*(Tfe-T0) - (3.45663e7 + 1.98865e9*xe)^0.5;
    g= -f;
  end

  function [dy] = ode_ex8_04(x,y,Cpf,Cpg,fc,DH,Ptot,R,S1,S2,U,W,NN20)
    Tf= y(1);
    Tg= y(2);
    NN2 = y(3);
    K1 = 1.789540e4*exp(-20800/(R*Tg));
    K2 = 2.57140e16*exp(-47400/(R*Tg));
    den=(2.598*NN20+2*NN2);
    pN2= Ptot*NN2/den;
    pH2= Ptot*3*NN2/den;
    pNH3 = Ptot*(2.23*NN20-2*NN2)/den;
    r = K1*pN2*(pH2^1.5)/pNH3 - K2*pNH3/(pH2^1.5);
    dy(3) = -fc*r;
    term1 = -U*S1*(Tg-Tf)/(W*Cpg);
    term2 = +(-DH)*S2/(W*Cpg)*(-dy(3));
    dy(1) = -U*S1*(Tg-Tf)/(W*Cpf);
    dy(2) = term1 + term2;
    dy=dy';
  end
```

Optimized decision variables: $x_{opt} = 7.51$ m; $T_{0,opt} = 704.1$ K

Maximum of objective function: 5.018.000 \$/year

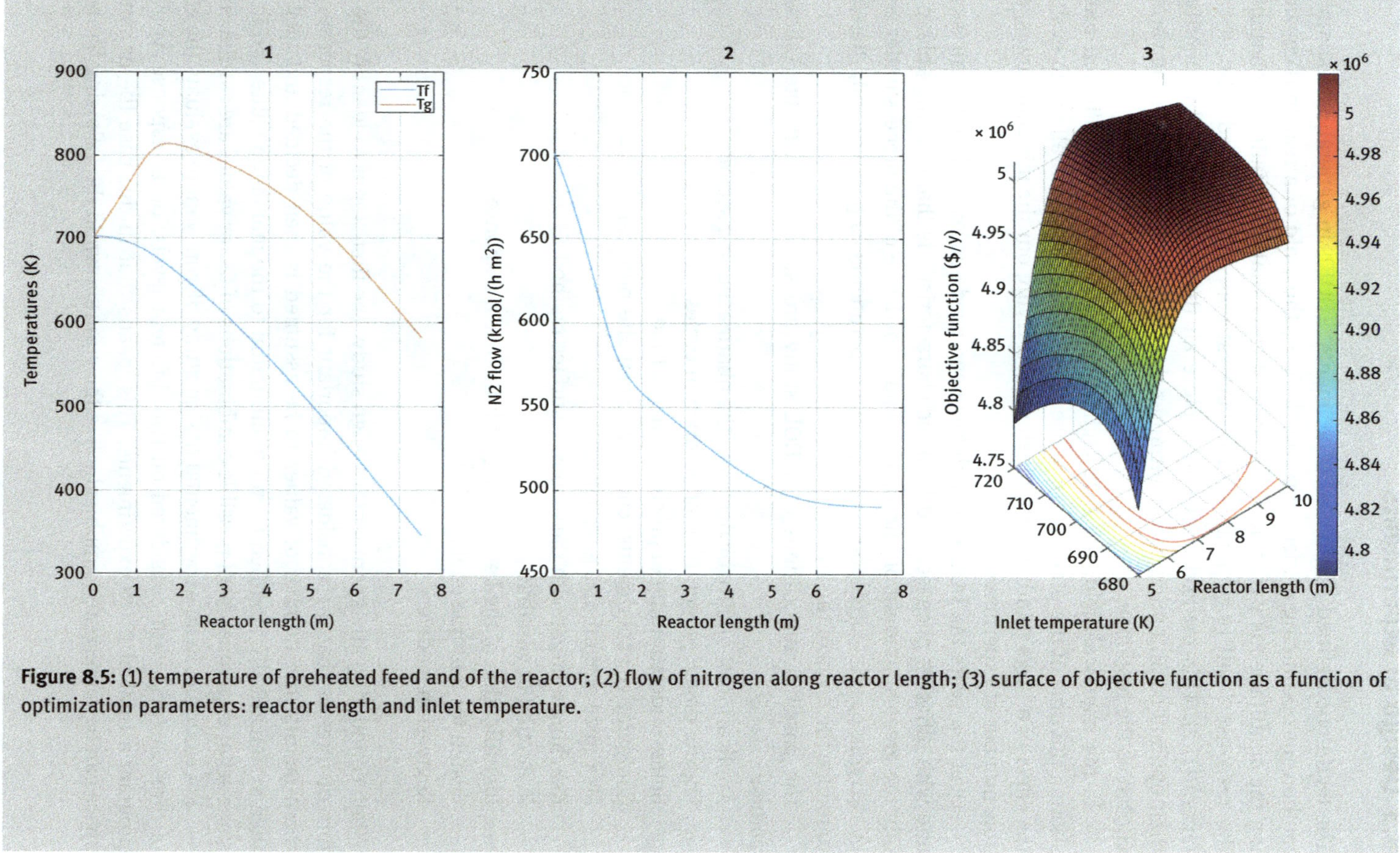

Figure 8.5: (1) temperature of preheated feed and of the reactor; (2) flow of nitrogen along reactor length; (3) surface of objective function as a function of optimization parameters: reactor length and inlet temperature.

8.3 Design of experiments

The technique defined “design of experiments” (DOE), in general, is a statistical method for optimizing a specific response (e.g., the yield of a reaction) considering the simultaneous variation of different factors that contribute to determining the response itself (e.g., temperature, catalyst concentration and ratio of reactants.) [1, 2]. This method allows to explore all the factors space by identifying their optimal combination that determines the best possible value for the variable chosen as the response.

In the chemical field, the DOE technique is currently a consolidated tool to shorten the time of the preliminary phase of reactions study as it allows to evaluate the effect of a large number of parameters by carrying out the minimum indispensable number of experiments.

Over the years the DOE approach has been used successful in different sectors and with different purposes and its basic characteristic is that of being able to obtain the greatest amount of information and knowledge of the process under study by carrying out, in correspondence, as few experiments as possible and, therefore, to reduce the costs of the study.

The possible applications of a DOE study are summarized in the following points:

- Application of quality by design of the pharmaceutical sector
- Shorter development times of production cycles
- Better understanding of processes in less time
- Comparative experiments (choice between alternatives)
- Maximize or minimize a response
- Reaching a target with the response surface method
- Reduction of variability
- Identification of the key factors that influence a response
- Search for multiple goals
- Regression models

When conducting an experimental design study, the first step is to identify the factors of interest that are considered important for the value of the response. For these factors, the number of values to be assumed is then decided and they are then systematically modified from the minimum to the maximum value and in the same set of experiments all possible combinations of factor values are carried out.

For example, in an experimental design study, if you want to evaluate the effect of three factors, each considered on two different levels, on a response, you can program a total of $2^3 = 8$ experiments. This means that in this case the parameter space can be represented graphically like the cube shown in the following figure.

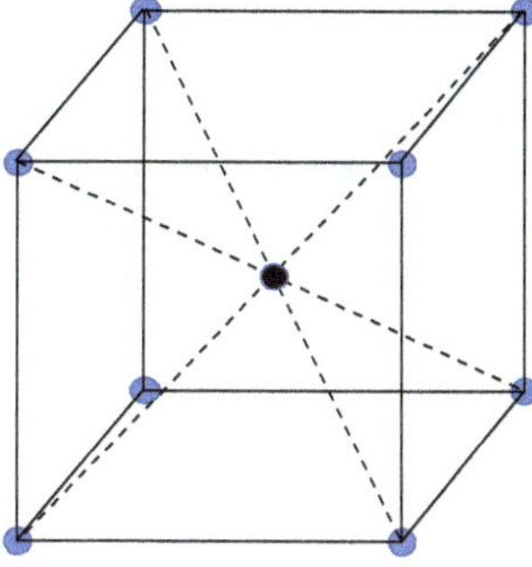

Figure 8.6: Schematic representation of a complete factorial design with three parameters.

The one illustrated is a case of a complete factorial design because all possible combinations of the three parameters are experimentally explored. Some very useful variants, compared to the experimentation scheme presented, consist of the following points:

- Carry out, in addition to the experiments already scheduled by the factorial design, the vertexes of the cube, an additional central experiment represented by the point in the center of the cube of figure 1.
- For more complete statistical information on experimental errors and variability, each experiment should be replicated more times (two or three usually).

The possibility of adopting a complete factorial design for an experimentation is not applicable, in practice, when the number of factors is very high. In this case, it is possible to use different techniques for reducing the total number of scheduled experiments with a modified factorial design that are called, for this reason, "reduced factorials." The reduction from the complete list of experiments to that of the reduced factorial occurs through the use of appropriate tables that can be found in specific texts [1].

Example 8.5 Maximizing process yield through DOE technique

It was decided to study the yield of a chemical process through the DOE. The response (yield) is studied by varying three factors (temperature, reagent concentration and heating time), each of which was varied on two levels coded as + and – (or +1 and −1 or high and low). Table 8.4 summarizes the setting of the DOE study factors.

Table 8.4: Factors and levels.

Factor	Code	Level −1	Level +1
Temperature (°C)	A	400	450
Concentration (% mol)	B	10	20
Heating time (s)	C	45	90

This configuration involves the execution of $2^3 = 8$ distinct experiments in which each possible combination of factors is explored. Furthermore, for gaining additional statistical information on the variability of the measurements it would be advisable to make replications (e.g., two or three) of each experiment. In the case of the present example, each experiment, correspondent to a given factors combination, was replicated three times for a total of 24 measurements. In the following table, the factors (expressed in encoded form) and the response (yield of the process) are reported.

Table 8.5: Codified factors and responses for DOE.

N. experiment	A (temperature)	B (concentration)	C (heating time)	Response (yield, %)
1	−1	−1	−1	66.63
2	−1	−1	−1	62.01
3	−1	−1	−1	57.85
4	1	−1	−1	77.25
5	1	−1	−1	70.33
6	1	−1	−1	67.73
7	−1	1	−1	50.25
8	−1	1	−1	59.95
9	−1	1	−1	56.05
10	1	1	−1	66.91
11	1	1	−1	70.16
12	1	1	−1	74.67
13	−1	−1	1	60.31
14	−1	−1	1	60.87
15	−1	−1	1	63.93
16	1	−1	1	69.98
17	1	−1	1	67.28
18	1	−1	1	67.54
19	−1	1	1	56.46
20	−1	1	1	58.03
21	−1	1	1	54.72
22	1	1	1	74.88
23	1	1	1	73.12
24	1	1	1	73.80

By using the data reported in Table 8.5 and the following polynomial model, evaluate the eight parameters and draw the residual plot and normal probability plot related to the residuals:

$$y = \beta_1 + \beta_2 A + \beta_3 B + \beta_4 C + \beta_5 AB + \beta_6 AC + \beta_7 BC + \beta_8 ABC$$

Matlab code for the solution of this example and results are reported as follows:

```
% example 8.5
clc, clear
global y g
global bopt

y= [ 66.63 62.01 57.85 77.25 70.33 67.73 50.25 59.95 56.05 . . .
   66.91 70.16 74.67 60.31 60.87 63.93 69.98 67.28 67.54 . . .
   56.46 58.03 54.72 74.88 73.12 73.80]';

g =[-1 -1 -1
   -1 -1 -1
   -1 -1 -1
   1 -1 -1
   1 -1 -1
   1 -1 -1
   -1 1 -1
   -1 1 -1
   -1 1 -1
   1  1 -1
   1  1 -1
   1  1 -1
   -1 -1 1
   -1 -1 1
   -1 -1 1
   1 -1  1
   1 -1  1
   1 -1  1
   -1 1  1
   -1 1  1
   -1 1  1
   1  1  1
```

```
  1  1  1
  1  1  1];

b0=[ 50 10 2 1 1 1 1 1 ];
lb=-1000*ones(1,8);
ub=+1000*ones(1,8);
options=optimoptions('lsqnonlin','Display','iter');
[b,resnorm,res]=lsqnonlin(@objfun_ex8_05,b0,lb,ub,options);

%% search for optimum yield
bopt=b;
F0=[ -1.1-0.510];
LB=[-1 -1 -1];
UB=[+1 +1 +1];
optionfmin=optimoptions('fmincon','Display','iter');
[F,fval]=fmincon(@fmax_ex8_05,F0,[],[],[],[],LB,UB,[],optionfmin);

%% output
disp(' ')
disp('-------------------------------')
disp(' Parameters ')
disp('-------------------------------')
fprintf(' Costant  %10.4f\n',b(1))
fprintf(' A  %10.4f\n',b(2))

printf(' B  %10.4f\n',b(3))
fprintf(' C  %10.4f\n',b(4))
fprintf(' AB  %10.4\n',b(5))
fprintf(' AC  %10.4f\n',b(6))
fprintf(' BC   %10.4f\n',b(7))
fprintf(' ABC%10.4f\n',b(8))
disp('-------------------------------')
disp(' ')
disp(' Factors')
disp('-------------------------------')
disp('')
fprintf('Max yield :    %10.4f\n ',abs(fval))
disp('')
fprintf('A      %5.1f  (450°C) \n',F(1))
fprintf('B      %5.1f  (10 perc.)     \n',F(2))
fprintf('C      %5.1f  (45 s)  \n',F(3))
disp('-------------------------------')

%% plots
figure(1)
```

```
subplot(1,2,1)
plot(res,'ro-')
grid
xlabel('Measurement index')
ylabel('Residual (maes-calc)')
title('Residuals plot')

subplot(1,2,2)
normplot(res)
xlabel('Residual')

figure(2)
interactionplot(y,g,'varnames',{'A','B','C'})
title('Effects plots')

function [fm] = fmax_ex8_05(F)
global bopt
A=F(1);
B=F(2);
C=F(3);
b=bopt;
fm=b(1) + b(2)*A + b(3)*B + b(4)*C + . . .
  b(5)*A.*B + b(6)*A.*C + b(7)*B.*C + . . .
  b(8)*A.*B.*C;
fm=-fm;
end

function [res] = objfun_ex8_05(b)
global y g
A=g(:,1);
B=g(:,2);
C=g(:,3);
ys=b(1) + b(2)*A + b(3)*B + b(4)*C + . . .
  b(5)*A.*B + b(6)*A.*C + b(7)*B.*C + . . .
  b(8)*A.*B.*C;
res=y-ys;
end

_____________
Parameters
_____________
  Costant    65.0296
  A         6.1079
  B         -0.9462
  C          0.0471
```

```
 AB        2.0654
 AC       -0.0846
 BC        1.0379
 ABC       0.6763
________________

Factors
________________

Max yield:  71.7700
A      1.0  (450 °C)
B     -1.0  (10 perc.)
C     -1.0  (45 s)
________________
```

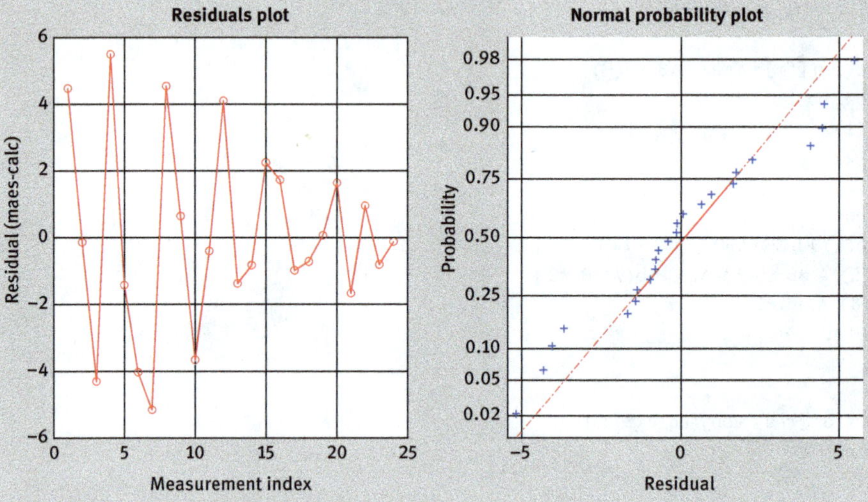

Figure 8.7: (1) Residual plot and (2) normal probability plot of residuals.

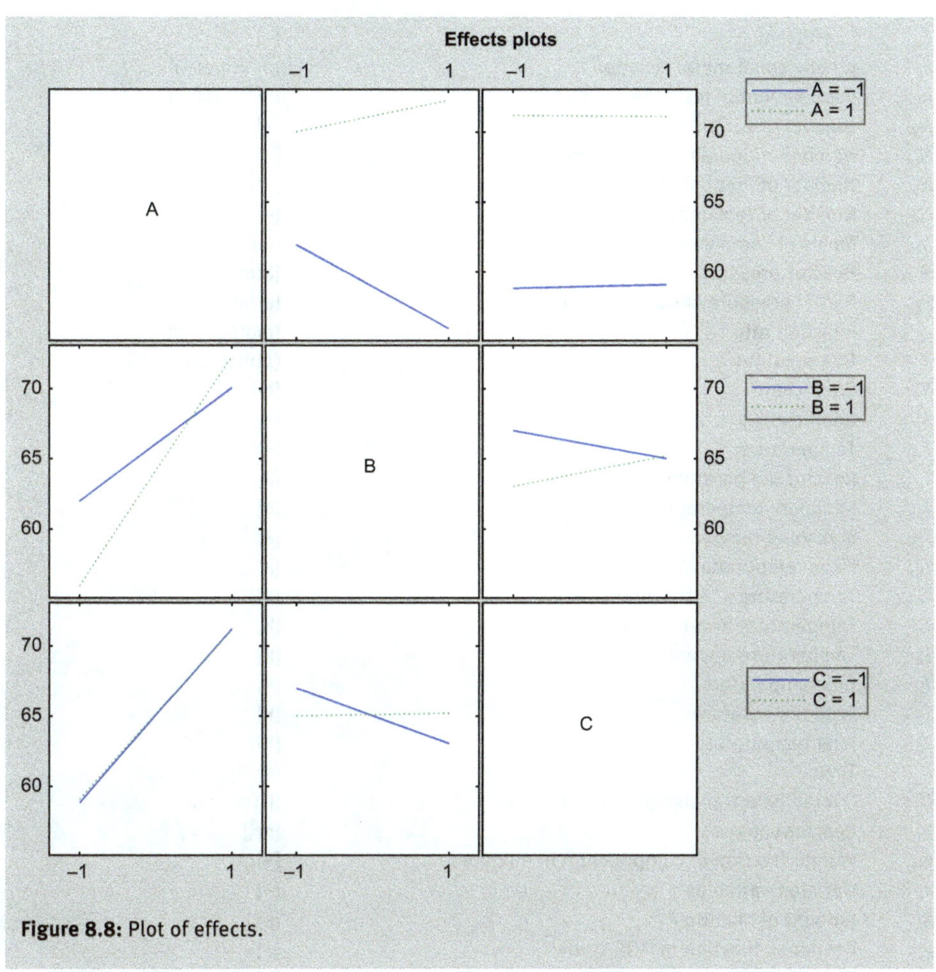

Figure 8.8: Plot of effects.

List of symbols

$A_{i,j}$	Coefficient matrix of linear system of equality or inequality constraints	[–]
b_j	Vector of known terms of linear system of equality or inequality constraints	[–]
C_i	Concentration of component *i*	[mol/m^3]
E_i	Activation energy of reaction *i*	[J/mol]
F_i	Molar feed flowrate of component *i*	[mol/s]
F_L	Lower bound of molar feed flowrate	[mol/s]
F_U	Upper bound of molar feed flowrate	[mol/s]

f_a	Coefficient	[–]
k_i	Kinetic constant for reaction *i*	$[m^3/(mol\ s)]$
k_{i0}	Preexponential factor for reaction *i*	$[m^3/(mol\ s)]$
N_V	Number of variables	[–]
N_{ec}	Number of equality constraints	[–]
N_{ic}	Number of inequality constraints	[–]
N_R	Number of reactions	[–]
n_i	Moles of component *i*	[mol]
$P_{reactor}$	Reactor pressure	[atm]
P_i	Partial pressure of component *i*	[atm]
r_k	Reaction rate	$[mol/(m^3 \cdot s)]$
R	Gas constant	$(atm\ m^3/(mol\ K)]$
S_1	Coefficient	[m]
S_2	Coefficient	$[m^2]$
T	Temperature	[K]
$T_{reactor}$	Reactor temperature	[K]
T_{min}	Minimum temperature	[K]
T_{max}	Maximum temperature	[K]
T_{fix}	Fixed temperature	[K]
T_i	Temperature at a fixed position	[K]
T_L	Temperature lower bound	[K]
T_U	Temperature upper bound	[K]
T_g	Gas temperature	[K]
T_f	Feed temperature	[K]
T_0	Inlet temperature	[K]
t	Time	[s]
U	Overall heat transfer coefficient	$[J/(m^2\ s\ K]$
V_R	Reactor volume	$[m^3]$
w_{Ai}	Weight fraction of component A in mixture *i*	[–]
x_i	Decision variables	[–]
Xi	Amount of mixture *i*	[kg]
y	Response function in DOE study	[–]
y_C	Yield of component c	[–]

Greek symbols

β_j	Adjustable parameter *j*	[–]
ΔH_j	Reaction enthalpy of reaction *j*	[J/mol]
$\nu_{i,j}$	Stoichiometric coefficient component *i* reaction *j*	[–]

References

[1] D.C. Montgomery. Design and Analysis of Experiments, (10th edition). Wiley: 2019.
[2] M.J. Anderson, P.J. Whitcomb. DOE simplified: Practical Tools for Effective Experimentation. Oregon: 2000.
[3] G.C. Onwubolu, B. V. Babu. New Optimization Techniques in Engineering. Springer: 2004.
[4] T.F. Edgar, D.M. Himmelblau, L.S. Lasdon. Optimization of Chemical Processes (2nd edition). McGraw-Hill: 2001.
[5] S.R. Upreti, K. Deb. Optimal design of an ammonia synthesis reactor using genetic algorithms. Computers and Chemical Engineering 1997, 21, 87–92.
[6] A. Murase, H.L. Roberts, A.O. Converse. Optimal thermal design of an autothermal ammonia synthesis reactor. Industrial & Engineering Chemistry Process Design and Development 1970, 9–4, 503–513.

Chapter 9
Case studies in chemical reaction engineering

9.1 Introduction

In this chapter some case studies are presented, taken from published papers, in which the modeling approach described in the previous chapters is applied. For simplicity, only the main equations will be reported. For further deepening, the readers are suggested to have a closer look to the cited literature, where all the details are reported.

9.2 Shrinking particle model

The present model was developed to simulate a special case study in chemical reaction engineering science: the reaction between a component dissolved in a liquid phase with a solid reactant. Even if several papers were published regarding film theory employed in the description of gas–liquid reactions, only few were devoted to the mentioned case.

The extended film theory can be successfully employed to describe the reactivity of soluble solid particles with liquids phase in which solid material is progressively dissolved. The dissolved specie diffuses then through the liquid film surrounding the particles giving place to chemical reaction in liquid phase (film and bulk). The particle shrinks as the dissolution proceeds and the film becomes thinner due to the decreasing particle size. A general mathematical model for conservation equations can be developed for the solid particle, the liquid film and the liquid bulk phase. The dynamic model is based on mass balances for the solid, film and bulk liquid and consists in a set of coupled parabolic partial differential equations (PDEs) and ordinary differential equations (ODEs).

The resulting equations set can be solved numerically by applying the method of lines and the following results can be obtained as model output: the shift of the reaction domain, being initially predominantly located in the liquid, film toward the liquid bulk during the progress of the solid–liquid process [1, 2].

The present case study has the objective to present an application of the film theory to reactive solids immersed in liquids. Such application regards numerical simulations of typical cases and parametric study.

https://doi.org/10.1515/9783110632927-009

9.2.1 Basic assumption of the model

The assumptions on which the shrinking particle/film model is based are as follows:
- The solid material is dispersed in a liquid phase; it dissolves and reacts with other components present in liquid.
- The solid particles are assumed all of the same size and the system is assumed isothermal.
- The chemical reactions are assumed as very fast and occur partially in the liquid film surrounding the solid particles and partially in the liquid bulk phase.
- The size of solid particles is reduced by dissolution and, consequently, also the liquid film around the particles becomes thinner. As the reaction progresses, the focus of the chemical transformation is shifted from the liquid film toward the liquid bulk.
- Perfect mixing is assumed so no concentration gradients appear in the liquid bulk.

The reaction system is described by the following relation:

$$r_i = \sum_{k=1}^{S} v_{ik} R_k \tag{9.1}$$

where R and r denote the reaction and generation rates, respectively, and ν_{ij} is the stoichiometric coefficient of component i in reaction k; S is the total number of chemical reactions in the system.

9.2.2 Mass balances and film thickness

In this section, the dimensionless form of the mass balance equations is reported. Details about the derivation can be found in the literature [1, 2].

In particular, for the liquid film surrounding the solid particle, the mass balance equation can be written as a function of the dimensionless film coordinate ($x = (r-R)/\delta$, where δ is the film thickness), trick useful to avoid solving a moving boundary PDE system:

$$\frac{dc_i}{dt} = \frac{D_i}{\delta^2}\left(\frac{d^2c_i}{dx^2} + \frac{s\delta}{R_p + \delta x}\frac{dc_i}{dx}\right) + r_i \tag{9.2}$$

with $0 \le x \le 1$. For relation (9.2) the following boundary conditions can be stated:
- $c_i = c^*_i$ (saturation of the solid components at the surface) at $x = 0$
- $dc_i/dx = 0$ (liquid-phase component at the surface) at $x = 0$
- $c_i = c'_i$ (bulk-phase conditions valid at the end of the film) $x = 1$

The film thickness can be calculated by using the standard correlation relating the Sherwood number (Sh) to the Reynolds (Re) and Schmidt (Sc) numbers, for example that of Wakao [3]:

$$\mathrm{Sh} = a' + b'\mathrm{Re}^{\alpha'}\mathrm{Sc}^{\beta'} \tag{9.3}$$

Extensive experimental studies of gas-phase systems have indicated that adequate values for the exponents in eq. (9.7) are $\alpha' = 1/3$ and $\beta' = 1/2$. By using the standard definition of Sh number and solving the film thickness ($\delta = D_i/k_{Li}$, $d/Sh = 2\ R/Sh$), the following result is obtained:

$$\delta = 2R_\mathrm{p}\left(2 + \left(\frac{\varepsilon}{\upsilon^3}\right)^{1/6}\left(\frac{\upsilon}{D_i}\right)^{1/3}(2R_\mathrm{p})^{2/3}\right)^{-1} \tag{9.4}$$

Two special cases are of particular interest: the first corresponds to $\varepsilon = 0$ (no energy is dissipated) and the second to vigorous turbulence (the term "2" is negligible compared to the second term in eq. (9.4)). The first limit case gives

$$\delta = R_\mathrm{p} \tag{9.5}$$

and for vigorous turbulence the expression is valid:

$$\delta = (2R_\mathrm{p})^{1/3}\left(\left(\frac{\varepsilon}{\upsilon^3}\right)^{1/6}\left(\frac{\upsilon}{D_i}\right)^{1/3}\right)^{-1} \tag{9.6}$$

For practical implementation, the particle radius should be related to the amount of solid substance (n_j). The amount of substance at a certain time is $n_j = n_\mathrm{P}\rho_\mathrm{P}V_\mathrm{P}/M_\mathrm{P}$, where n_P is the total number of particles in the system. At the beginning of the reaction, the amount of substance is instead $n_{0j} = n_P\rho_\mathrm{P}V_{0\mathrm{P}}/M_\mathrm{P}$. Considering that for a general particle geometry, $V_\mathrm{P}/V_{\mathrm{OP}} = (R_\mathrm{P}/R_{\mathrm{PO}})^{s+1}$ we can obtain the relationship:

$$\frac{n_j}{n_{0j}} = \left(\frac{R_\mathrm{p}}{R_{\mathrm{p}0}}\right)^{s+1} \tag{9.7}$$

is obtained. By combining expressions (9.7) and (9.4) we obtain the final operative equation for the film thickness:

$$\delta = R_{\mathrm{p}0}\left(\frac{n_j}{n_{0j}}\right)^{1/(s+1)}\left(1 + \left(\frac{\varepsilon}{\upsilon^3}\right)^{1/6}\left(\frac{\upsilon}{D_i}\right)^{1/3}\left(\frac{R_{p0}{}^2}{2}\right)^{1/3}\left(\frac{n_j}{n_{0j}}\right)^{2/(3(s+1))}\right)^{-1} \tag{9.8}$$

From eq. (9.8) is evident that the film thickness diminishes from the initial value to zero as the solid reactant (j) is consumed.

The bulk-phase mass balance can be written as follows:

$$\frac{dc'_i}{dt} = r'_i + N'_i a_0 (R_p/R_{p0})^s \tag{9.9}$$

Substituting eq. (9.7) into (9.9) we finally obtain

$$\frac{dc'_i}{dt} = r'_i + N'_i a_0 (n_j/n_{0j})^{s/(s+1)} \tag{9.10}$$

The flux at the film–bulk interface is calculated from

$$N'_i = -\left(\frac{D_i}{\delta}\right)\left(\frac{dc_i}{dx}\right) \quad at\, x = 1 \tag{9.11}$$

Finally, the solid phase mass balances are given by

$$\frac{dn'_j}{dt} = \left(\frac{D_j a_0 V_L}{\delta}\right)\left(\frac{R_p}{R_{p0}}\right)^s \frac{dc_j}{dx}\bigg|_{x=0} \tag{9.12}$$

for a general particle geometry. After inserting the relationship between the amount of substance and the radius, the balance becomes as follows:

$$\frac{dn'_j}{dt} = \left(\frac{D_j a_0 V_L}{\delta}\right)\left(\frac{n_j}{n_{0j}}\right)^{s/(s+1)} \frac{dc_j}{dx}\bigg|_{x=0} \tag{9.13}$$

9.2.3 Dimensionless model and solution procedure

For the implementation purposes, model equations presented in the previous sections have been transformed into dimensionless forms and the full details of the derivation are reported elsewhere [1, 2]. The model obtained was then constituted by few dimensionless equations:

$$\frac{dy_i}{d\theta} = \frac{D_i \tau}{R_{p0}{}^2 (\delta/R_{p0})^2}\left(\frac{d^2 y_i}{dx^2} + \frac{s(\delta/R_{p0})}{R_p/R_{p0} + (\delta/R_{p0})x}\frac{dy_i}{dx}\right) + r_i \tau/c_0 \tag{9.14}$$

where $0 \le x \le 1$, $y_i = c_i/c_0$, $\theta = t/\tau$, τ = characteristic (arbitrary) time and $D_i\tau/R_0{}^2$ is the first dimensionless number defined in the present work.

A dimensionless film thickness is obtained as follows,

$$\delta/R_{p0} = \left(\frac{n_j}{n_{0j}}\right)^{1/(s+1)}\left(1 + \alpha\left(\frac{n_j}{n_{0j}}\right)^{2/(3(s+1))}\right)^{-1} \tag{9.15}$$

where α is the second dimensionless number. Concerning the liquid bulk, the final form of the dimensionless equation becomes

$$\frac{dy'_i}{d\theta} = \frac{r'_i\tau}{c_0} - \frac{D_i a_0 \tau / R_{p0}}{\delta / R_{p0}} \frac{dy_i}{dx} \left(\frac{n_j}{n_{0j}}\right)^{s/(s+1)} \tag{9.16}$$

where $D_i\, a_0\tau/R_0$ is the third dimensionless number. The final form becomes

$$\frac{dn'_j/n_{0j}}{d\theta} = \left(\frac{D_j a_0 \tau / R_{p0}}{\delta / R_{p0}}\right)\left(\frac{c_0 V_L}{n_{0j}}\right)\left(\frac{n_j}{n_{0j}}\right)^{s/(s+1)} \frac{dy_j}{dx}\Big|_{x=0} \tag{9.17}$$

The calculations procedure can be summarized as follows:

1) The dimensionless film thickness is solved explicitly from eq. (9.15).
2) The dimensionless concentration profiles in the liquid film are solved from eq. (9.14).
3) The dimensionless concentrations in the liquid bulk are obtained from eq. (9.16).
4) The dimensionless amount of solid substance is calculated from eq. (9.17).

The model consists of coupled ODEs and parabolic PDEs, the ODEs representing the liquid bulk-phase components and the solid-phase PDEs representing the components in the liquid film. The model equations along with the initial and boundary conditions were solved numerically by using MATLAB as modeling tool. The PDE system was solved with the built-in numerical method of lines, in particular using the second-order centered finite difference formulae (CFDM) for the spatial derivatives, with 40 discretization points along the film, thus the film equations were transformed to a set of ODEs, an initial value problem.

9.2.4 Contribution analysis

The role of the film (F) and bulk phases (B) in the process can be evaluated by considering the corresponding contributions. The contribution analysis can be sketched as follows for, respectively, the film and bulk contributions. Details are given in the cited literature [1, 2]

$$\lambda_{kF} = \frac{\int_0^{V_F} R_k dV_F}{\int_0^{V_F} R_k dV_F + R'_k V_L} \tag{9.18}$$

$$\lambda_{kB} = \frac{R'_k V_L}{\int_0^{V_F} R_k dV_F + R'_k V_L} \tag{9.19}$$

9.2.5 Simulation example

The considered example describes a system, in which the reaction between a dissolving component (B) and a component in the liquid phase (A), reacting in both the film and the bulk phases, takes place. Saturation prevails in the liquid phase in the immediate vicinity of the solid–liquid interface. The reaction scheme is displayed:

1. $B_S \rightarrow B$ Dissolution
2. $A + B \rightarrow C$ $r_1 = k \cdot nc_A(x) \cdot c_B(x)$
3. $A + B \rightarrow C$ $r_2 = k \cdot nc_A \cdot c_B$

A qualitative analysis of the concentration profiles is given in Figure 9.1, where the dimensionless concentration profiles along time and film thickness for the three components are reported.

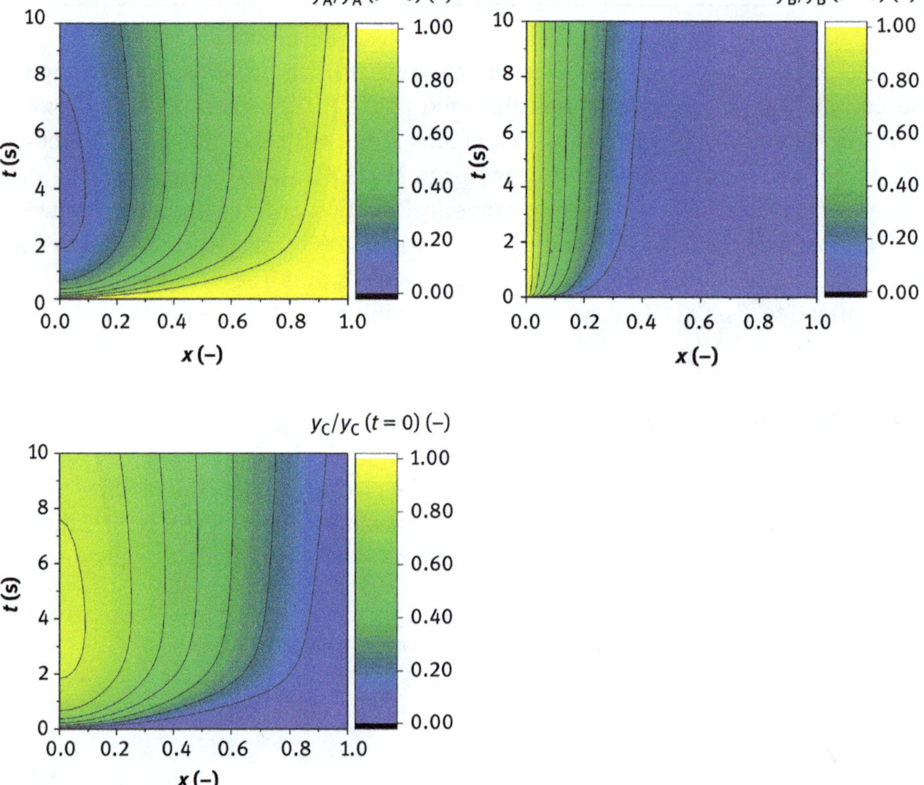

Figure 9.1: Contour plots for the three components along time and film thickness. Figure inspired by Salmi et al. [1].

Components A and C show similar profiles, but they are opposite in absolute values, being the reagent and the product. Component B which is assumed to have an initial concentration in the film equal to zero dissolves and reacts simultaneously, showing a decreasing profile from the particle surface to the film along the time. These aspects can be better revealed from Figure 9.2.

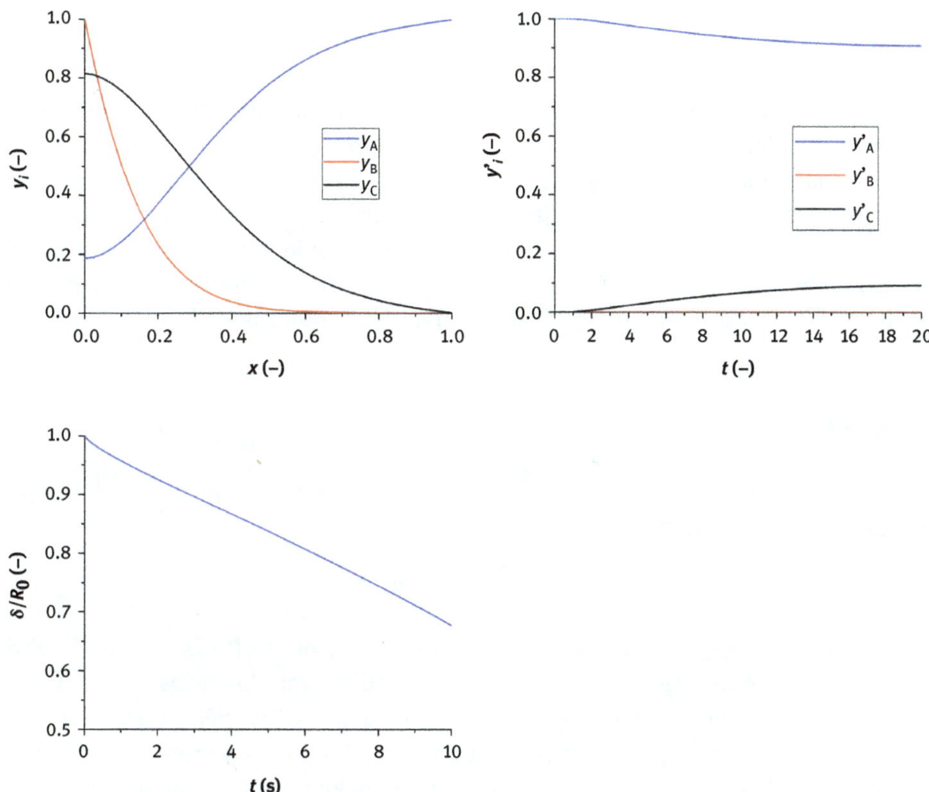

Figure 9.2: Dimensionless concentration profiles: (a) at 1 s in the film phase, (b) in the bulk phase and (c) film thickness evolution with reaction time. Figure inspired by Salmi et al. [1].

As the reaction proceeds, the model predicts a reasonable evolution with time of the dimensionless component concentrations in both bulk and film, showing a gradual dissolution of component B. Finally, as the reaction proceeds, the film thickness gradually gets thinner because of the solid particle dissolution.

9.3 Falling film reactors modeling

9.3.1 Introduction

The technology of falling film reactors represents a valid solution for the industrial conduction of sulfonation reactions, or in general reactions being highly exothermic. The reaction occurs inside a liquid reactive film flowing downward on the internal reactor wall. In this way two goals can be attained: (i) an optimal heat exchange can be realized toward the external jacket, (ii) the scale-up of such reactors is quite straightforward as it consists mainly on increasing the tubes numbers. The main problem of this reactor setup is then its modeling.

The mathematical model of such reactor should consider both the heat and mass transfer phenomena in both gas and liquid phases. Moreover, by considering that normally the liquid film flows under a mixed laminar/turbulent regime, the system increases in complexity, because it is necessary to implement the mass diffusion and heat transfer contributes also in the radial coordinate of the reactor.

9.3.2 Reactor model

The development of the falling film reactor model is reported in the literature [4] and here briefly summarized, for a reaction:

$$A_{(l)} + S_{(g)} \rightarrow P_{(l)}$$

The main assumption of the model is that the liquid phase flows under a mixed laminar/turbulent regime, while the gas phase is turbulent. The mass and heat balance equations were written in a dimensionless form, by performing variable changes, in order to keep into account for a moving boundary between gas and liquid that is not fixed as the liquid film can vary in thickness along reactor axis.

In steady-state conditions, mass balance for gas and liquid phases for each component are reported:

$$c_{S,G} = c_{S,G,\text{feed}}\left(1 - \frac{x_A}{R_{S/A}}\right) \tag{9.20}$$

$$u_L \frac{\partial c_{S,L}}{\partial z} = +k_G a_G\left(c_{S,G} - m c_{S,L}\big|_{y=0}\right) + \frac{\partial}{\partial r_f}\left[(D_S + D_T)\frac{\partial c_{SO_3,L}}{\partial r_f}\right] + \nu_S r \tag{9.21}$$

$$u_L \frac{\partial c_{A,L}}{\partial z} = \frac{\partial}{\partial r_f}\left[(D_A + D_T)\frac{\partial c_{A,L}}{\partial r_f}\right] + \nu_A r \tag{9.22}$$

With $0 \le z \le L$ (reactor axial coordinate) and $\delta_G \le r_f \le \delta_G + \delta_L$ (liquid film coordinate). With the progress of the reaction, the liquid viscosity increases thus the liquid slows down leading to an increase of the liquid film thickness. For this reason, the discretization of the liquid film changes along the axial coordinate. This aspect was treated by a variable change, introducing dimensionless coordinates. A new radial coordinate has been introduced (y) that changes linearly with the film radius as in relation (9.23):

$$r_f = \delta_G + \delta_L y^*, \quad \partial r_f = \delta_L \partial y^* \tag{9.23}$$

Hence, the mass balances can be written as in eqs. (9.24)–(9.25):

$$\frac{u_L}{L}\frac{\partial c_{S,L}}{\partial \zeta} = +k_G a_G (c_{S,G} - m c_{S,L}|_{y=0}) + \frac{1}{\delta_L}\frac{\partial}{\partial y^*}\left(\frac{D_S + D_T}{\delta_L}\frac{\partial c_{S,L}}{\partial y^*}\right) + \nu_S r \tag{9.24}$$

$$\frac{u_L}{L}\frac{\partial c_{A,L}}{\partial \zeta} = \frac{1}{\delta_L}\frac{\partial}{\partial y^*}\left(\frac{D_A + D_T}{\delta_L}\frac{\partial c_{A,L}}{\partial y^*}\right) + \nu_A r \tag{9.25}$$

For what concerns energy conservation, the steady-state heat balance for the gas and liquid phases are reported in eqs. (9.26)–(9.27), written in dimensionless form:

$$\frac{u_G}{L}\frac{\partial \rho_G c_{p,G} T_G}{\partial \zeta} = \frac{h_G}{\delta_L}(T_L|_{y=1} - T_G) \tag{9.26}$$

$$\frac{u_L}{L}\frac{\partial \rho_L c_{p,L} T_L}{\partial \zeta} = -\frac{1}{\delta_L}\frac{\partial}{\partial y^*}\left(\frac{-k_L}{\delta_L}\frac{\partial T_L}{\partial y^*}\right) + (-\Delta_r H) r \tag{9.27}$$

The set of boundary conditions (BCs) needed to solve the problem, follow the strategy reported:

- plug-flow behavior at the pipe inlet
- continuity at the pipe exit
- symmetry at the pipe center
- continuity at the pipe wall
- continuity equations at gas–liquid interface

For the complete definition of the model, in addition to PDEs and Boundary Conditions (B.C.)s, a set of relations must be adopted for the calculation of other parameters of the system, for example, mass transfer specific area, liquid film thickness, linear velocity of gas and liquid, liquid density and viscosity. For more details, the readers should refer to the paper by Russo et al. [4].

9.3.3 Modeling strategy

The PDE system has been solved with the method of lines that is present in MATLAB since R2017b release. The derivatives along the radial coordinate of the liquid film were calculated with the second-order CFDM with 20 discretization points while the axial coordinate of the reactor (gas and liquid) has been solved with ode15s algorithm.

9.3.4 Simulation example

The developed model allows the simulation of pilot and industrial size reactors. As an example, in Figure 9.3 the results of a simulation are reported for sulfonation reaction, in terms of SO_3 in both gas and liquid phases (the profiles are referred to the gas–liquid interphase). This concentration decreases along the pipe length as the reaction proceeds. The reaction appears very fast and exothermic and occurs almost completely in the first 20% of the reactor length. Another interesting simulation result is the increase of liquid film thickness along the length of the reactor. In the first part, due to a temperature increase, leading to a viscosity decrease, the film gets slightly thinner. Then, as the reaction proceeds, conversion increases, so the viscosity of the liquid phase increases, thus the film becomes thicker.

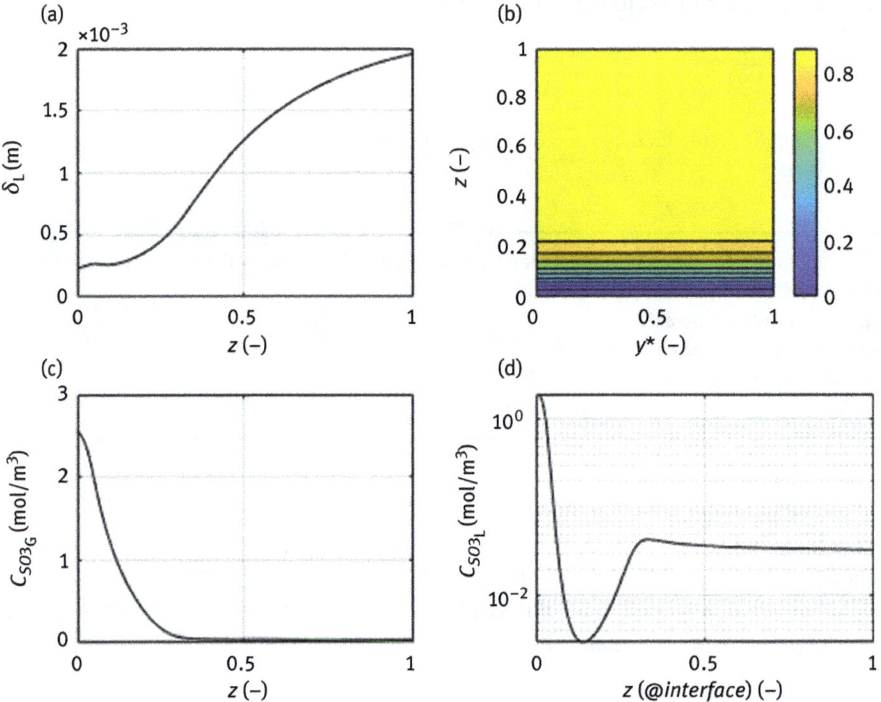

Figure 9.3: Simulated profiles related to the mass balance solution of a sulfonation reaction. (a) Liquid film thickness versus dimensionless axial coordinate. (b) Substrate conversion contour plot as a function of both axial and radial coordinates. (c) SO_3 concentration in the gas phase along dimensionless axial coordinate. (d) SO_3 concentration in the liquid phase along dimensionless axial at gas–liquid interface. Figure inspired by Russo et al. [4].

From Figure 9.3, SO_3 continues dissolving and reacting after a first steep decrease in the concentration profile due to the gas–liquid mass transfer that is lower than the reaction rate. No significant radial gradients are present, due to the high turbulent liquid diffusivity of the system given by the high gas flowrate.

In Figure 9.4, the results of the heat balance are reported: as the conversion increases, the temperature increases too, and after the reaction is completed, the dominating thermal effect is related to the cooling jacket that cool down both liquid and gas phases. As expected, due to the balance between the high exothermicity of the reaction and the heat exchanged with the reactor jacket, the temperature shows a maximum.

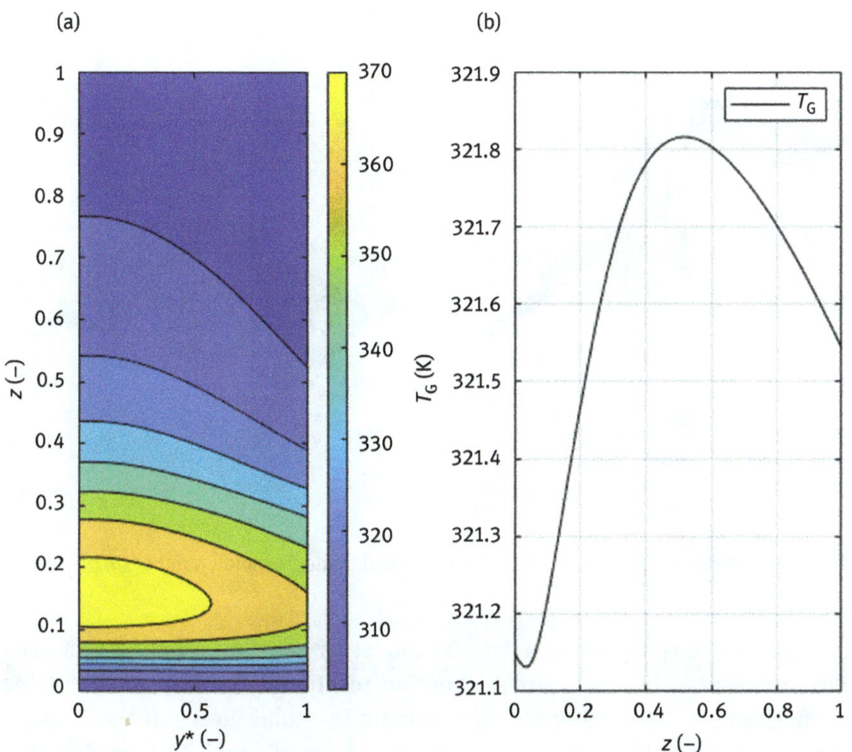

Figure 9.4: Simulated profiles related to the heat balance solution. (a) Liquid temperature contour plot versus dimensionless axial and radial coordinates. (b) Gas temperature profile. Figure inspired by Russo et al. [4].

Gas-phase temperature change is almost negligible, due to both the low thermal conductivity and the short residence time of the gas phase.

The gas velocity is a key factor that strongly influences the trend of the concentration profiles along the film thickness direction. The value of gas velocity can determine radial film concentration gradient or can lead to a well-mixed film in the radial direction for which the gradient is absent. The results obtained are reported in Figure 9.5, verifying that when a laminar contribution is present, the Reynolds number is below 20, indicating a laminar flow regime.

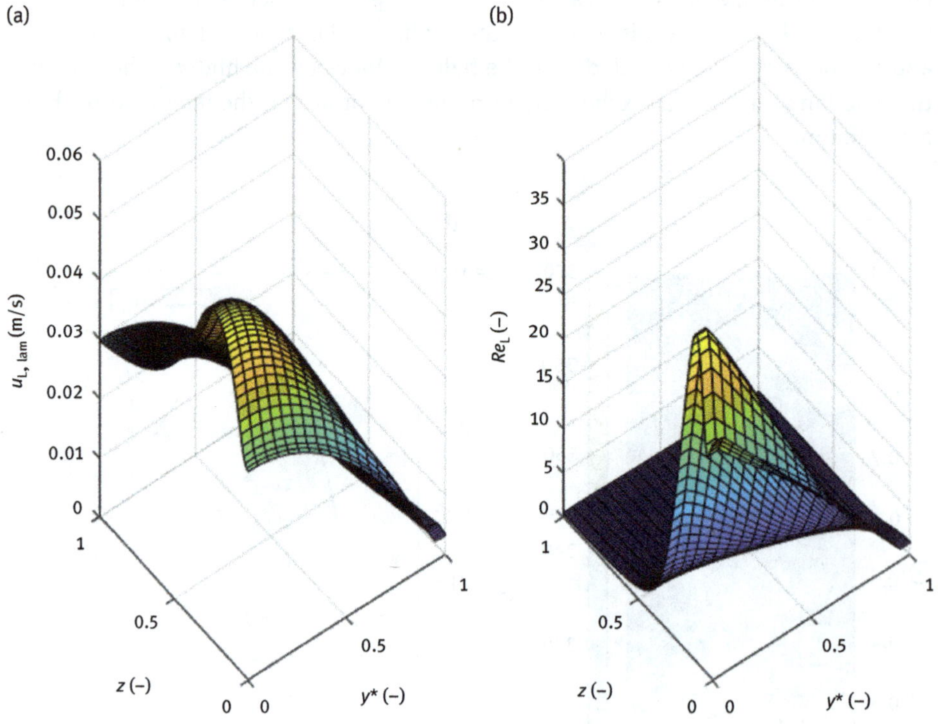

Figure 9.5: Simulation results of laminar flow velocity (a) and liquid Reynolds number (b).

As we have seen in this case study, a detailed model for gas–liquid falling-film reactor is able to describe the setup for sulfonation reactions. The key aspect of the model is the discretization of the solution domain in radial direction while along the reactor axis the resulting differential equations are of the ODE type. In other words, this strategy allows to transform PDE equation into ODE one. In Matlab environment, available built-in functions were used to discretize the film thickness using the numerical method of lines and implementing finite difference approximation for spatial derivatives.

At last, film thickness variation with conversion is accurately described, indicating that a large increase of the liquid thickness is predictable when the system reaches full conversion. This fact has consequences also on the gas velocity profiles, allowing to calculate variable Reynolds number for both gas and liquid phases. The liquid velocity profile was rather turbulent, because of the high shear stress of the gas phase. As a conclusion, the laminar contribution can be considered negligible in the adopted operation conditions. Therefore, the model can be considered general, able to predict also cases where the gas phase velocity is lower.

9.4 Microreactors modeling

9.4.1 Introduction

Microreactors are considered as one of the most powerful tools in the investigation of the intrinsic kinetics of gas-phase reactions, as the catalyst layers in microreactor channels are very thin (20–30 micrometer), and the heat transfer characteristics of microreactors are excellent. For the mentioned reasons, intraparticle mass transfer limitations should be suppressed, thus the microreactors are normally simulated using ideal approaches.

Therefore, it was recently demonstrated that for fast reactions, it is not possible to neglect intra wash-coat diffusion limitation to correctly retrieve the intrinsic kinetics of a given system [5–7]. Thus, a clear need of more sophisticated model emerged and were developed. In the present section, the main efforts on the mathematical modelling of microreactors with MATLAB are reported.

9.4.2 Modeling approach

The fluid-phase mass balance equation for the wash coated microreactor is reported in eq. (9.28). The model is dynamic, including convective and dispersion fluxes

$$\frac{\partial C_{i,\mathrm{F}}}{\partial t} = -u_{\mathrm{F}} \cdot \frac{\partial C_{i,\mathrm{F}}}{\partial z} + D_{z,\mathrm{F}} \cdot \frac{\partial^2 C_{i,\mathrm{F}}}{\partial z^2} + D_{\mathrm{r,F}} \cdot \left(\frac{\partial^2 C_{i,\mathrm{F}}}{\partial r^2} + \frac{1}{r} \cdot \frac{\partial C_{i,\mathrm{F}}}{\partial r} \right) \tag{9.28}$$

Solid-phase mass balances are reported in eq. (9.29):

$$\frac{\partial C_{i,\mathrm{s}}}{\partial t} = \frac{D_{\mathrm{e},i}}{\varepsilon_{\mathrm{p}}} \cdot \left(\frac{\partial^2 C_{i,\mathrm{s}}}{\partial {r_{\mathrm{p}}}^2} + \frac{s}{r_{\mathrm{p}}} \cdot \frac{\partial C_{i,\mathrm{s}}}{\partial r_{\mathrm{p}}} \right) + \frac{\sum (v_{i,j} \cdot r_j)}{\varepsilon_{\mathrm{p}}} \tag{9.29}$$

The boundary conditions are defined as follows:
- plug-flow behavior at the reactor inlet
- Danckwerts' closed boundary condition at the reactor outlet
- symmetry at the pipe center
- continuity condition at the fluid–solid interface

9.4.3 Numerical methods

The used discretization method used a central difference approximation for axial and radial derivatives (numerical method of lines). Both axial and radial coordinates were approximated with first- and second-order central finite differences, with 50 discretization points for the reactor length and 10 for the reactor radius, using MATLAB software.

9.4.4 Modeling results of specific cases

The first case presented deals with ethylene oxide synthesis using Ag/Al_2O_3 catalyst [5]. The authors modeled a wash coated microreactor to investigate the intrinsic kinetics of the reaction. Experimental data were taken from the literature and the model was applied with success to simulate all the available data with parameter estimation analysis, fact only possible including the intraparticle diffusion limitations in the model as, with a preliminary approach, it was not possible to simulate simultaneously all the collected data [8].

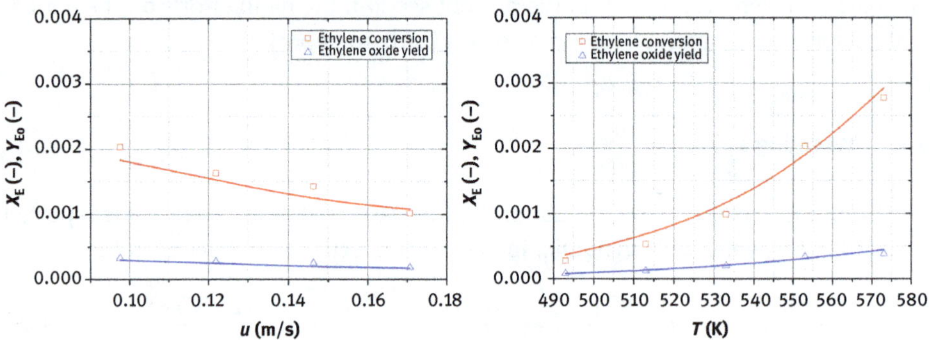

Figure 9.6: Experimental data and modeling results of wash coated microreactor for ethylene oxide synthesis. (a) Ethylene conversion and ethylene oxide yield as a function of the fluid velocity at T = 553 K. (b) Ethylene conversion and ethylene oxide yield as a function of temperature. Superficial gas velocity = 0.1 m/s. Figure inspired by Russo et al. [5].

As can be seen, the reaction is strongly dependent on both fluid velocity and temperature.

The same model was applied to ethanol oxidation using golden nanoparticles as catalyst [6] with remarkable good results, also in this case only possible as intraparticle diffusion limitation was considered (Figure 9.7), allowing to properly describe both reactants and products distribution profiles with temperature.

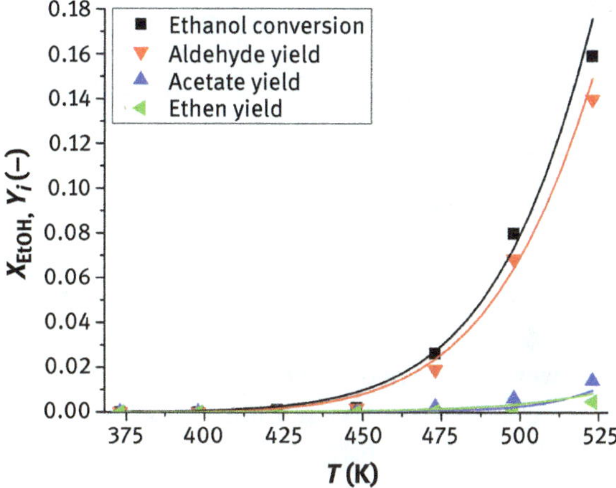

Figure 9.7: Ethanol conversion and product yields (aldehyde; acetate; ethene) as a function of temperature, fixing a temperature rump of 1 °C/min and a feed flowrate of 25 mL/min. Figure inspired by Behravesh et al. [6].

9.5 Reactive chromatography modeling

9.5.1 Introduction

Reactive chromatography is a hot topic in chemical reaction engineering science. A chromatographic reactor is a single unit where the chemical reaction and chromatographic separation are possible in the same unit [9]. It represents a clear case of how it is possible to intensify a chemical process. The principle is that the solid packed in the reactor must act both as catalyst and stationary phase, thus the solid phase must show high reactivity and selective interaction to reactants/products.

This special reactive unit is particularly well suited for reversible reactions as the separation of the products is responsible for the increase of the reactant conversion, shifting the chemical equilibrium to the products side.

The modeling of chromatographic reactors was recently deepened by Russo et al. [9], proposing a rigorous dynamic adsorptive chromatographic reactor model (DACR), a fluid–solid model that can be considered a real improvement of the modeling approaches reported in the literature. The model is based on the description of the chemistry and physics in detail, focusing the attention on all the possible mass transfer steps appearing in the chromatographic reactor. The main equations and general solutions are reported in this section.

9.5.2 Chromatographic reactor model

The DACR was written in a time-dependent form, to simulate the evolution of the concentration profiles along the reactor bed. The liquid phase flows in the axial z coordinate. The mass balance equation for the fluid phase is reported in eq. (9.30), considering both axial dispersion and liquid film mass transfer:

$$\frac{\partial C_{i,\mathrm{B}}}{\partial t} = -u_L \frac{\partial C_{i,\mathrm{B}}}{\partial z} + D_z \frac{\partial^2 C_{i,\mathrm{B}}}{\partial z^2} - \frac{k_m a}{\varepsilon'} \left(C_{i,B} - C_{i,\mathrm{L}}\big|_{rp=Rp} \right) \tag{9.30}$$

Fluid velocity and total concentration were considered constants as one of the reactants is pulsed as small quantity to a stream of the other reactant, implying a non-influent change in the overall composition and a slightly low influence on the fluid velocity.

Two boundary conditions were needed, which imply the following:

- Pulse function at the reactor inlet for the component added by pulse to the continuous stream:

$$C_{i,\mathrm{B}}\big|_{z=0} = \begin{cases} t \le t_{\mathrm{inj}}: C_{i,\mathrm{feed}} \\ t > t_{\mathrm{inj}}: 0 \end{cases}, \quad t_{\mathrm{inj}} = \frac{V_{\mathrm{loop}}}{\dot{V}} \tag{9.31}$$

- Plug-flow condition at the reactor inlet for the other components.
- Zero derivative for every component at the reactor outlet (Danckwerts' closed boundary condition).

The pulsed-stream B.Cs take into account that a small quantity of a reactant (i) flows in the reactor with an injection time that is proportional to the volume of the injection loop and the volumetric flowrate of the main stream (component j).

The intraparticle mass balance equation was written to include both liquid inside pores and surface diffusion phenomena, describing adsorption by calculating the diffusion path in a solid particle as the sum of two parallel contributions: (i) the porous diffusion, depending on the particle porosity and tortuosity; (ii) surface diffusion, depending on the interaction between each component and the solid surface.

The mentioned assumptions lead to the following mass balance equations, eq. (9.32):

$$\left[(1-b_i)\varepsilon_{\mathrm{p}} + b_i\right] \frac{\partial C_{i,\mathrm{L}}}{\partial t} = (1-\varepsilon_{\mathrm{p}})v_i r + \varepsilon_{\mathrm{p}} D_{\mathrm{e},i} \frac{1}{r_{\mathrm{p}}^{\mathrm{s}}} \frac{\partial}{\partial r_{\mathrm{p}}} \left(r_{\mathrm{p}}^{\mathrm{s}} \frac{\partial C_{i,\mathrm{L}}}{\partial r_{\mathrm{p}}} \right) + (1-\varepsilon_{\mathrm{p}}) D_{S,i} \frac{1}{r_{\mathrm{p}}^{\mathrm{s}}} \frac{\partial}{\partial r_{\mathrm{p}}} \left(r_{\mathrm{p}}^{\mathrm{s}} b_i \frac{\partial C_{i,\mathrm{L}}}{\partial r_{\mathrm{p}}} \right)$$

$$C_{i,S} = b_i C_{i,\mathrm{L}} \tag{9.32}$$

The adsorption isotherm can be approximated to be linear as the pulsed component becomes diluted in the continuous stream. This assumption is certainly wrong in the case of a fixed-bed adsorptive reactor, where the two reactants are fed simultaneously at the inlet of the reactor, thus the system cannot be considered diluted and the implementation of a dedicated adsorption isotherm is needed, see references [10–13].

Four sets of boundary conditions are needed:
- At the reactor inlet, steady-state conditions were assumed to prevail.
- The concentration derivative is zero for all the components at the reactor outlet.
- Symmetry conditions can be imposed to the particle center.
- A continuity equation needed is applied to the catalyst surface ($r_p = R_p$).

The fluid–solid mass transfer coefficient can be calculated from existing correlations, for example, Dweivedi and Upadhyay correlation, originally published and tested for values obtained using a packed-bed reactor, eq. (9.33) [14]:

$$k_m = \frac{1.1062}{\varepsilon_\mathrm{P}}\left(\frac{D_\mathrm{P} u_F \rho_\mathrm{F}}{\mu_\mathrm{F}}\right)^{\alpha''} u_F \left(\frac{\mu_\mathrm{F}}{\rho_\mathrm{F} D}\right)^{-2/3}, \alpha'' = \begin{cases} -0.72, \mathrm{Re_P} \le 10 \\ -0.4069, \mathrm{Re_P} > 10 \end{cases} \tag{9.33}$$

9.5.3 Modeling approach

The PDE system describing the chromatographic reactor was solved with the method of lines implemented in Matlab. The axial coordinate of the reactor was solved with an asymmetrical backward finite difference approximation, defining 150 grid points transformed by log10. The reason of the logarithm transformation was to achieve a better resolution of the chromatographic peaks in the initial part of the reactor where the peaks are approaching the δ function of Dirac due to the injection. The radial coordinate of the particle was approximated by a second-order central finite difference method with 20 discretization points.

9.5.4 Example of application

Methyl formate hydrolysis was taken as test reaction, leading to formic acid and methanol as products. In our previous investigation, all the physical, kinetic and thermodynamic parameters were retrieved either from correlations or estimated. We suggest the reader to check our paper for further details [9]. Several simulations were conducted, but in this section, the attention will be focused only on the main results.

The influence of the volumetric flowrate on the chromatographic reactor output is demonstrated in Figure 9.8(a). An increase of the flowrate suppresses the axial dispersion, thus Pe increases, leading to sharper peaks. Moreover, the flowrate increase corresponds to a decrease in the residence time, leading to two main effects:
- shift of the chromatographic peaks at lower time
- decrease of the overall conversion

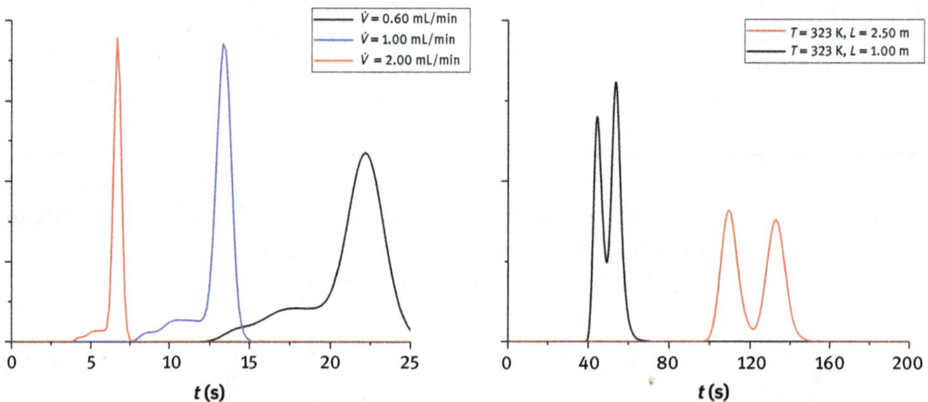

Figure 9.8: DACR simulation output. (a) Effect of the volumetric flowrate. (b) Effect of reactor length. Figure inspired by Russo et al. [9].

The influence of the reactor length on the column performance shows interesting results (Figure 9.8(b)), as by increasing L the system is characterized by higher residence time that brings higher conversion and shifts of the chromatographic peaks, because the column is characterized by a higher number of theoretical plates. This result demonstrates that by using a chromatographic reactor, it is possible to obtain full conversion also for equilibrium reactions, as the reactants and products are separated in the column. As demonstrated, the products are separated, and no other purification unit is needed.

List of symbols

A	Surface area	[m^2]
a	Surface area-to-volume ratio	[m^2/m^3]
a', b'	Coefficients in the correlation for the Sherwood number	[–]
b	Adsorption parameter	[m^3/mol]
c	Concentration	[mol/m^3]
c_p	Specific heat	[J/(kg K)]
d	Particle diameter (=2 R)	[m]
D	Diffusion coefficient	[m^2/s]

D_e	Effective diffusion coefficient	$[m^2/s]$
D_r	Radial dispersion coefficient	$[m^2/s]$
D_z	Axial dispersion coefficient	$[m^2/s]$
k	Reaction rate constant	$[m^3/(mol \cdot s)]$
k_G	Mass transfer coefficient	[m/s]
k_L	Mass transfer coefficient	[m/s]
k_m	Mass transfer coefficient	[m/s]
L	Reactor length	[m]
m	Partition coefficient	[–]
M	Molar mass	[kg/mol]
N	Diffusion flux	$[mol/(m^2 \cdot s)]$
n	Amount of substance	[mol]
R, R'	Reaction rates	$[mol/(m^3 \cdot s)]$
$R_{S/A}$	Molar ratio	[–]
r, r'	Generation rate	$[mol/(m^3 \cdot s)]$
r	Pipe radial coordinate	[m]
r_p	Particle radial coordinate	[m]
R_p	Particle radius	[m]
s	Shape factor	[–]
T	Temperature	[K]
t	Time	[s]
u	Flow velocity	[m/s]
V	Volume	$[m^3]$
x	Dimensionless film coordinate	[–]
X	Conversion degree	[–]
y	Dimensionless concentration	[–]
y^*	Dimensionless film thickness coordinate	[–]
Y	Yield	[–]
z	Reactor axial coordinate	[m]

Greek letters

α	Dimensionless number	[–]
α', β'	Exponent in the correlation for Sherwood number	[–]
α''	Exponent in the correlation for Dweivedi and Upadhyay correlation	[–]
δ	Film thickness	[m]
$\Delta_r H$	Reaction enthalpy	[J/mol]
ε	Dissipated energy	[W/kg]
ε'	Solid/fluid volumetric ratio	[–]
ε_p	Particle void degree	[–]
ζ	Dimensionless reactor axial coordinate	[–]
θ	Dimensionless time	[–]
κ	Contribution	[mol/s]
λ	Normalized contribution	[–]
μ	Dynamic viscosity	[Pa s]
ν	Stoichiometric coefficient	[–]

υ	Kinematic viscosity	$[m^2/s]$
ρ	Density	$[kg/m^3]$
τ	Characteristic reaction time	[s]

Dimensionless numbers

Re	Reynolds number	[–]
Re_p	Reynolds particle number	[–]
Sc	Schmidt number	[–]
Sh	Sherwood number	[–]

Subscripts and superscripts

B	Bulk phase
f	Fluid
F	Film
G	Gas phase
i	Liquid-phase and general component index
inj	Injection
j	Solid-phase component index
k	Reaction rate index
L	Liquid
P	Particle
S	Solid
T	Turbulent
0	Initial quantity
*	Saturated state

References

[1] T. Salmi, V. Russo, C. Carletti, T. Kilpiö, R. Tesser, D. Murzin, T. Westerlund, H. Grénman. Application of film theory on the reactions of solid particles with liquids: Shrinking particles with changing liquid films. Chemical Engineering Science 2017, 160, 161–170.

[2] V. Russo, T. Salmi, C.A. Carletti, D.Yu.Murzin, T. Westerlund, R. Tesser, H. Grénman. Application of an Extended Shrinking Film Model to limestone dissolution. Industrial & Engineering Chemistry Research 2017, 56(45), 13254–13261.

[3] N. Wakao. Recent Analysis of Chemically Reacting Systems (Ed. Doraiswamy, L.K.). Wiley Eastern: 1984.

[4] V. Russo, A. Milicia, M. Di Serio, R. Tesser. Falling film reactor modelling for sulfonation reactions. Chemical Engineering Journal 2019, 377, 120464.

[5] V. Russo, T. Kilpiö, J. Hernandez Carucci, M. Di Serio, T. Salmi. Modeling of microreactors for ethylene epoxidation and total oxidation. Chemical Engineering Science 2015, 134, 563–571.

[6] E. Behravesh, T. Kilpiö, V. Russo, K. Eränen, T. Salmi. Experimental and modelling study of partial oxidation of ethanol in a micro-reactor using gold nanoparticles as the catalyst. Chemical Engineering Science 2018, 176, 421–428.

[7] Y. Khan, T. Kilpiö, M. Marin, V. Russo, J. Lehtonen, R. Karinen, T. Salmi. Modelling of a microreactor for the partial oxidation of 1-butanol on a titania supported gold catalyst. Chemical Engineering Science 2020, 221, 115695.

[8] J. Carucci, V. Halonen, K. Eränen, J. Wärnå, S. Ojala, M. Huuhtanen, R. Keiski, T. Salmi. Ethylene oxide formation in a microreactor: From qualitative kinetics to detailed modelling. Industrial & Engineering Chemistry Research 2010, 49, 10897–10907.

[9] V. Russo, R. Tesser, C. Rossano, R. Vitiello, R. Turco, T. Salmi, M. Di Serio. Chromatographic rector modelling. Chemical Engineering Journal 2019, 377, 119692.

[10] V. Russo, R. Tesser, M. Trifuoggi, M. Giugni, M. Di Serio. A dynamic intraparticle model for fluid-solid adsorption kinetics. Computers and Chemical Engineering 2015, 74, 66–74.

[11] V. Russo, R. Tesser, D. Masiello, M. Trifuoggi, M. Di Serio. Further verification of adsorption dynamic intraparticle model (ADIM) for fluid-solid adsorption kinetics in batch reactors. Chemical Engineering Journal 2016, 283, 1197–1202.

[12] V. Russo, D. Masiello, M. Trifuoggi, M. Di Serio, R. Tesser. Design of an adsorption column for methylene blue abatement over silica: From batch to continuous modelling. Chemical Engineering Journal 2016, 302, 287–295.

[13] V. Russo, M. Trifuoggi, M. Di Serio, R. Tesser. Fluid-solid adsorption in batch and continuous processing: A review and insights on the modelling. Chemical Engineering & Technology 2017, 40(5), 799–820.

[14] P.N. Dwivedl, S.N. Upadhyay. Particle-fluid mass transfer in fixed and fluidized beds. Industrial & Engineering Chemistry Process Design and Development 1977, 16, 157–165.

Word Index

https://doi.org/10.1515/9783110632927-010

www.ingramcontent.com/pod-product-compliance
Lightning Source LLC
Chambersburg PA
CBHW061342210326
41598CB00035B/5863

* 9 7 8 3 1 1 0 6 3 2 1 9 4 *